Teach® Yourself

Get Started in
Pig Keeping
Tony York

For UK order enquiries: please contact Bookpoint Ltd,
130 Milton Park, Abingdon, Oxon OX14 4SB.
Telephone: +44 (0) 1235 827720. Fax: +44 (0) 1235 400454.
Lines are open 09.00–17.00, Monday to Saturday, with a 24-hour
message answering service. Details about our titles and how to
order are available at www.teachyourself.com

For USA order enquiries: please contact McGraw-Hill Customer
Services, PO Box 545, Blacklick, OH 43004-0545, USA.
Telephone: 1-800-722-4726. Fax: 1-614-755-5645.

For Canada order enquiries: please contact McGraw-Hill Ryerson
Ltd, 300 Water St, Whitby, Ontario L1N 9B6, Canada. Telephone:
905 430 5000. Fax: 905 430 5020.

Long renowned as the authoritative source for self-guided learning –
with more than 50 million copies sold worldwide – the **Teach Yourself**
series includes over 500 titles in the fields of languages, crafts, hobbies,
business, computing and education.

British Library Cataloguing in Publication Data: a catalogue record
for this title is available from the British Library.

Library of Congress Catalog Card Number: on file.

First published in UK 2007 by Hodder Education,
part of Hachette UK, 338 Euston Road, London NW1 3BH.

First published in US 2007 by The McGraw-Hill Companies, Inc.

This edition published 2010.

Previously published as *Teach Yourself Keeping Pigs*.

The **Teach Yourself** name is a registered trade mark of
Hodder Headline.

Typeset by MPS Limited, a Macmillan Company.

Printed in Great Britain for Hodder Education, an Hachette UK
Company, 338 Euston Road, London NW1 3BH, by CPI Cox &
Wyman, Reading, Berkshire RG1 8EX.

The publisher has used its best endeavours to ensure that the URLs
for external websites referred to in this book are correct and active
at the time of going to press. However, the publisher and the author
have no responsibility for the websites and can make no guarantee
that a site will remain live or that the content will remain relevant,
decent or appropriate.

Hachette UK's policy is to use papers that are natural, renewable
and recyclable products and made from wood grown in sustainable
forests. The logging and manufacturing processes are expected to
conform to the environmental regulations of the country of origin.

Impression number   10 9 8 7 6 5 4 3 2 1
Year   2014 2013 2012 2011 2010

Contents

Meet the author

Welcome to *Get Started in Pig Keeping*!

'Happiness is pig-shaped' is a phrase that I have used from the very first moment I started keeping pigs in the 1970s and I can think of no better phrase to use now. Unlike a lot of things that might cause you to think that you are in your own private heaven, you do not have to wait for days, weeks or years for the feeling of complete happiness to engulf you – it is instant. From the very first moment you select your first piglets the smile on your face will be permanent. Small-scale pig keeping is not so much a way of life but much more a way to happiness as the contentment of being a pig keeper engulfs you. Whether you are a lawyer, doctor, nurse, housewife or even a high-powered captain of industry, the world of pig keeping will give you something extra. Pig keeping offers rewards far greater than just quality food for the family or the satisfaction that comes when your first litter of piglets is born. You will quickly find that as you open the entrance to your pig pen the door of your workplace instantly closes behind you.

In the world of pig keeping, one of the most significant changes that has taken place over the last few years is the remarkable increase in the number of people, from all walks of life and socio-economic backgrounds, who are exploring the possibility of keeping a few pigs. Maybe the very fact that you are reading this book means that you may soon be joining this happy throng of people about to experience the benefits of keeping pigs.

Pat yourself on the back if you intend keeping a few pigs for the table as a means of ensuring that you and your family have the benefit of knowing that what you eat has been fed and reared to the standards that you feel happy and content with.

Pat yourself on the back if you are considering keeping one of our
rare or traditional breeds for breeding, in order to help ensure
that these wonderful breeds do not become extinct.

Pat yourself on the back if you are one of the increasing number of
people who realize that there must be more to life than many of the
constant pressures associated with modern-day living. Wanting to
do something that relieves some of these work pressures and offers
a better quality of life is as good a reason for keeping a few pigs as
you could want.

Pat yourself on the back if you are one of those many people who
realize that a pig, more than almost any other animal, gives more
than it takes and you are just keeping a pig for the pure joy of it.

If you fit into any or all of these categories 'pat yourself on the back'
because, put quite simply, you have made the right decision.

Get Started in Pig Keeping will be your bookcase mentor, providing
you with all the information you need to start keeping pigs.
With years of keeping pigs myself, and years of experience running
'One Day Pig Keeping Courses', I have been able to listen to what
must be almost every problem facing a newcomer to pig keeping.

Tony York

Only got a minute?

Small-scale pig keeping is rewarding in so many ways that even just by thinking about starting to keep pigs you will begin to feel the blood racing and a million questions rushing through your mind. Each chapter in this book will reward your enthusiasm and you will find the following:

► Provided you have enough land anyone can keep pigs.

► Registering to keep a pig is very simple but you **must not** keep a pig before you have followed the correct procedures.

► A small area of rough scrubland or woodland is ideal for a pig.

► Pigs are remarkably clean animals and do not need a great deal of space.

► Contrary to the opinion held by many people pigs do not smell if kept and fed correctly.

- Provided you follow some basic rules pigs are not consistently noisy animals.

- Pigs do not need constant attention unlike many other animals. As well as needing to be fed and watered, cows need milking; sheep need shearing; chickens need securing from foxes; dogs need walking, and while all these animals are adorable you will find pigs need very little apart from food, water and kindness.

- Attending a short course on keeping pigs and taking advice from the many pig clubs, associations and societies before you embark on your adventure will save you money and heartache.

- Unlike many new ventures, small-scale pig keeping need not be a drain on your finances and it can be quite the reverse.

- Pig keeping can be rewarding both financially and emotionally as well as producing 'a mouth-watering eating experience'.

5 Only got five minutes?

All too often people form the impression that pigs are animals to be feared, kept away from children, are difficult to keep, expensive, not 'cost effective' and worst of all that they will need the security of Alcatraz to keep them contained. If you think this then you are wrong. This book dispels these myths and, I hope, makes you realize how much of a pleasure keeping pigs can be. While reading this book you will discover the following:

▶ Getting the right pig for you and what you plan to do with your pig is very important. If the pig is right for you, then you will be right for the pig.
▶ Provided that you get the right breed of pig, at the right age, from the right breeder and then treat and feed it correctly, you will find that they do not have a bad bone in their body.
▶ Pigs are not born nasty – they are made nasty. Look very carefully at the person you buy the pig from and see how their pigs have been kept. Buy from a reputable breeder and avoid buying from markets where you will get no provenance.
▶ Pigs kept for the table not only more than pay for themselves but also give you and your family a truly 'mouth-watering eating experience'.
▶ Providing a secure pen, paddock or home for your pig is really very simple. Pigs are intelligent animals and, just like you, will not want to escape or run away if they are fed and cared for correctly. Barbed wire at the bottom of the fencing will stop them burrowing out.
▶ The pig is genetically very close to a human so you must learn that many of the problems that some pig keepers come across are the result of 'bad or poor parenting'. Treat your pig as well as you treat your children and if they are of 'teenager age' be just as firm. Pigs are hierarchal and respect both kindness and authority.

If you have never kept any form of livestock before and perhaps have never even kept a domestic pet, then your move into keeping pigs might seem daunting in the extreme. If so, put away any doubts or fears you may have and this book will help you understand the following:

▶ The pig is probably the easiest of all the animals to keep and certainly one of the easiest to train in order to make sure it behaves and reacts in the way you want it to.
▶ You will soon prefer your pigs to a lot of humans for they will help you in your new role, not hinder you. They will bring out all the good in you and force you to ask the question 'Why didn't I do this years ago?'
▶ You are only required to be yourself. There is no need for you to have a farming background or be in any way pretentious.

If you are concerned about the commitment, time and effort you might have to put into your new venture then all you need to do is simply understand that by taking the slowly, slowly and step-by-step approach to keeping pigs that is advocated in this book you can do as much or as little as you choose. Pigs are very hardy animals and will happily spend most of their time outside provided they have a warm, dry house to sleep in. You can also choose when you want to keep them if you are rearing to provide food for the table, i.e. just in the summer/only once a year. If you buy weaners to fatten for pork you only have the pigs for 17 weeks so there's no long-term commitment.

How often or for how long you keep pigs is up to you, your circumstances and your family's needs. Pig keeping is a world of flexibility with all the advantages that this brings with it, but beware, it can become addictive!

10 Only got ten minutes?

Thinking about and even talking about keeping pigs is one thing but actually doing it is another. You need to 'have at hand' not just enthusiasm (for although it is a valuable and important quality to possess it will rise and fall with each stumbling step you take!) but you need above all to have guidance and knowledge.

It would be wonderful if all new pig keepers could spend their first year in the company of an experienced pig keeper simply watching, learning, helping and enjoying but that is neither possible nor practical. Instead, you must turn to the written word. Contained in this book is more than 30 years of practical pig keeping experience combined with knowledge gained from over 20 years lecturing and teaching new pig keepers from all parts of the world. No matter where you look or what page you turn to you will find the answers you are looking for:

What do I need to buy before I get my pig? Apart from somewhere for it to sleep in and a feed, trough and water bowl the next most important thing to have by you is a thermometer.

Which is the best breed for me? There are many to choose from and all are different not just in appearance, size and temperament but also in potential profitability.

Why can't I get quality pork and quality bacon from the same pig at the same age? Quality pork comes at a younger age than quality bacon.

What can I do if my pig appears sick? Take the pig's temperature and assess the degree of ill health then examine thoroughly and, if necessary, report your findings to your vet.

How much room do I need? Very little room at all, in fact the 'smallholder's pig' used to be kept in a pig sty often only 10 ft by 12 ft with an outside area as a toilet area.

What do I need to do and get organized before I get my first pig?
Learn as much as possible about keeping pigs as you can then
make sure you have suitable housing, a suitable well-fenced area
for keeping pigs in, feeding troughs and water troughs plus a CPH
(County Parish Holding) number and a Movement licence.

**I am not a vet so will I be able to cope or will it cost me a fortune if
my pig is ever sick?** Pigs do not get sick very often if you feed them
correctly while looking after them in a proper manner. Pigs are easily
treated and respond very quickly to treatment if they do become ill
and you can often treat them easily. If you are just buying piglets
to raise for pork you will only have them to look after for around
17 weeks so there is not a lot of time for them to become ill.

What can I feed my pig? Do not feed meat, meat-related products,
gravy or anything that has been in contact with meat or anything
else that has come from any kitchen. Feed your pigs specially
manufactured pig pellets/nuts prepared by the major food
manufacturers – you can get them from your local agricultural
merchants – plus fruit, vegetables and bread, etc.

What rules and regulations must I observe? Your pig must be
clearly marked (metal tag, tattoo or 'slap-mark') for traceability
and you must record movement on and off your premises as well
as recording any drugs given to your pig.

What should I look for when buying a pig? It is often more
important to look at the person you are buying from and how their
pigs are kept than anything else. If you intend to show then make
sure they meet all the breed standards.

How much do pigs cost? This is an impossible question to answer
by giving an 'across the board' price. It is best to visit the pigs or
piglets, select the ones you want and then ask how much.

Can I kill, butcher and eat my own pig? Legally yes. But it is not
advised because it is always better for a licensed abattoir to kill
your pig and a professional licensed butcher to prepare your meat.

If not then you are prevented by law from selling your meat or giving it to anyone else to eat.

What size pig ark do I need? Many people end up buying an ark that is too big for the pig so always get the smallest ark suitable for what you need. Pigs are very clean animals and will never dirty their bed but if they have an ark that is too big they will turn it into an en-suite!

How many times a day do I need to feed my pig? Always feed at least twice a day (morning and evening) because pigs are basically browsers and it is not good to give them vast amounts of food all at once. You need to see them twice a day to make sure that everything is alright.

Do I need to worm my pigs? Yes. Pigs should be wormed at eight weeks and again every six months.

Do I need to vaccinate my pigs? Vaccinate breeding stock two weeks before farrowing and vaccinate against Erysipelas at eight–nine weeks of age and again 14 days later.

Do pigs get on with other animals? Pigs are very sociable and get on well with dogs, cats, goats, cattle, sheep and, to many people's surprise, even horses. Although to start with horses are very wary of something they don't know and also the strange smell, once they have got used to it they are fine together.

Are pigs expensive? Definitely 'no'. When you take into consideration that you can pay many hundreds of pounds for a pedigree puppy, the same and sometimes more for a pure-bred kitten, thousands of pounds for an ostrich and even more for an alpaca then piglets that barely reach three figures are very cheap.

Where should I go to buy my pigs? Always go to a reputable breeder. All the recognized breeds have a Breed Society and they will give you a list of recognized breeders to contact.

I have young children so is it wise to have pigs? Provided that you are a good 'parent' to both the pig and your children then all will be well. If you allow a young child in with a big pig before they have got to know each other you are taking an unnecessary risk. If you allow any of your children, no matter what their ages, to hit your pig, throw stones at it or physically abuse it, the pig is bound to react badly. This aside, pigs and children get on very well indeed with a pig being more trustworthy than a lot of dogs as far as children are concerned. You should always insist, however, that everyone washes their hands after being in with the pigs.

Are pigs noisy? Pigs are only noisy at feed times because it is their way of saying 'Hurry up I'm hungry' but the noise dies down as soon as they get their food. However, if you give food (even treats such as apples or bananas, etc.) to your pigs every time you go to see them they will soon become noisy whenever they see someone coming because they think they will be getting food.

Are pigs smelly? Provided you feed them the right things and practice good hygienic husbandry you will not suffer from bad odours or stinking pigs.

Can I put my pet pig on a lead and take it for walks? Yes but you must first get a 'walking licence' from your local Trading Standards Office (Animal Health) department who will ask you to tell them what your 'walking route' will be. If this route goes anywhere near a pig farm or near where other pigs are kept then it is likely to be refused.

How many litters can a pig have a year? Theoretically they could have two and a half litters a year which is the aim of some commercial units but this is just like using a sow as a baby machine and I would never recommend it if you have any feelings for either the mother or her piglets. Weaning at 56 days, and then giving the mother three to six weeks to recover, means that you are realistically looking at two litters a year.

As a small-scale pig keeper who am I responsible to? The main body who will oversee your pig-keeping activities will be your local Trading Standards Office (Animal Health) department. They will give you all the help and advice as far as paperwork is concerned, the rules and regulations plus the legal side of pig keeping.

Introduction

The world of pigs, especially rare and minority breeds, is a place that many people would like to visit but where so many fear to tread. *Get Started in Pig Keeping* has been written in order to replace fear with reality, and apprehension with confidence while at the same time opening the doors to a whole new experience that will undoubtedly enhance your life. Used correctly this book will do just that.

If you are expecting this book to be full of technical jargon, charts, growth performance tables and feed analysis you will be disappointed. It has been written and structured in such a way that everyone interested in keeping pigs on a small scale can benefit rather than just catering for a few specialists. From the very first chapter you will be able to grasp the importance of placing yourself on an equal footing with the pig if you truly want all the benefits that small-scale pig keeping can give you. It is not a book you will want to put down, but once read, it is a book you can easily refer to again and again whenever the need arises. Each chapter deals with a specific area of interest to the new pig keeper but while, as a whole, this book encompasses everything you will need to know, you have a lot more than just a reference book. It has been written to enthuse and encourage as well as simply teach.

If you take each chapter in order you will be guided gently through the process of keeping pigs right from those early days when you were pondering whether or not to take the first steps right through to those magic moments when you achieve the total satisfaction that keeping pigs can bring you. On the other hand, once you have started your new adventure, you will benefit constantly from referring back to specific chapters and advice as you begin to move forward and tread new ground.

Don't expect the impossible and simply read this book as if it were a 'once read – all known' document because it is much more than that and you should treat it as your friend. The more you get to know it and the more you refer to it then the more will be the benefits and, just like the pig itself, the more you give to it the greater will be your gain.

Image credits

1

The starting point

In this chapter you will learn:
- *sensible reasons why you might want to keep a pig*
- *how to assess your suitability as a pig keeper*
- *the questions you need to ask yourself before embarking on your new venture.*

Why keep pigs?

It would be easy to answer such a question by saying, quite simply, that it is because pigs are the most adoring, intelligent, rewarding and giving animals you will ever come across but there are two problems with such a glib statement. Firstly, if you have ever had anything at all to do with pigs before you read this book then you already know that this is a fact. Secondly, if keeping pigs is going to be a new adventure for you then it is only by keeping them you will realize that, if anything, this is an understatement rather than some wild 'pig mad' exaggeration. Let us therefore start by looking more closely at the benefits you are likely to gain from becoming a small-scale pig keeper.

Winston Churchill once said, 'A dog looks up to you, a cat looks down on you but a pig treats you as an equal.' How right he was. When you look a pig in the eye it looks back at you with an assured confidence that clearly reflects its determination and ability to learn almost as much about you by the way you behave as you intend to learn about the pig. Don't ever underestimate the pig's ability to assess people, situations and their actions for this is the very quality that enables them to be one of the great animal 'givers' rather than just 'takers'. All of us, no matter what our individual background might be, much prefer 'givers' to 'takers' and this is one of the first benefits you will get when you become a pig keeper for the first time. It does not matter whether you are just keeping a pig as a 'pet', thinking of breeding to help save some of our rare and traditional pig breeds or if you are planning to become a 'hobby farmer' or smallholder keeping two or three pigs so you can guarantee a better quality of meat for your family and friends – the result is the same. Pigs will give you the benefit of a better quality of life as a result of their ability to 'give' you more than they take.

> ## Insight: Pigs teach you something new every day
> When I started out myself I made the mistake of thinking I would know everything within a month or two – over 30 years later I am still learning!

POSSIBLE REASONS FOR KEEPING A PIG

There are three main reasons why you may be considering keeping a pig and although you may, over a period of time, want to change the type of pig keeping you are involved in, it is important that you decide, well in advance, on your initial aims and ambitions. If you do not make this a priority you are liable to get the wrong breed of pig to start with, causing you untold problems right from the start. You have the following choices but no matter which route you take, if you choose a rare pig breed then you are also playing your part in helping to save these important traditional breeds for future generations:

1 *To keep three or four pigs to produce quality meat for your family and friends so that you have a 'mouth-watering eating experience' as a result of knowing how your pig has been kept, what it has been fed and how the meat has been prepared. A definite benefit to any new pig keeper who wants to get*

*away from commercially fed and reared pork, ham, gammon
and bacon. To travel down this route you start by buying
three (the optimum number) weaner piglets, at around eight
weeks of age, from a recognized breeder and keep them until
they are ready for pork at 26–32 weeks of age or bacon/
gammon/ham at 40–50 weeks of age.*

Insight: Cheap piglets

Having two of the first four piglets I bought die after four
weeks taught me that buying piglets cheap was a false
economy and buying before they were properly weaned was
even worse.

2   *To embark upon small-scale pig breeding thereby producing
your own piglets either to sell or to fatten yourselves for the
freezer. This will save you from having to buy weaner piglets
from another breeder and give you the added benefit of being
involved in increasing breed numbers thus playing your part
in helping to protect and save some of our rare and traditional
pig breeds. You can either take the 'quick start' route by
getting an adult pig that is due to farrow (an 'in-pig' gilt or
sow) or start with weaner piglets and take the slower route
towards producing your own 'in-pig' gilt that will farrow
and produce piglets for you after you have had her for about
12 or 13 months. Both routes have their advantages and
disadvantages that we will discuss in detail later in this book.
Breeding your own stock has the specific benefit experienced
by anyone involved in breeding any animal (human or
otherwise) – that magic moment of the birth itself with the
excitement of watching these tiny animals grow and develop
as a result of your involvement. It is impossible for even the
most sceptical of new pig keepers not to be moved by the sight
of a newly born piglet, dwarfed by its mother, struggling to
gain the strength and balance that will enable it to lock on to
one of its mother's teats and get its first vital meal outside of
its mother's womb. You may not shed a tear but if you do we
will not be surprised.*

3 *To simply keep a pig as a 'pet'. It is no more wrong to keep a pig as a 'pet' than it is to keep one for breeding or for meat provided you always keep your pig in the right environment and in the correct conditions. You may want to keep a specialist 'pet' pig breed that does not grow as large as the traditional breeds or you may just want a larger rare breed pig that will share its life and your land with you for many years to come. We will discuss the various breeds and your options in a later chapter but whatever breed you ultimately decide to keep you will have the benefit of a 'friend for life' who will give you hours of pleasure over many years. It is not unusual for a pig to live for between 10 and 15 years and you will soon find that they are better than any 'anti-stress' medication you could ever be given. Your pig will take you to a new, almost unimaginable world where the pressures of modern day living simply do not exist.*

Over a period of time you may, of course, embark on any combination of the above or even all three together but it is wise and sensible to start slowly and choose just one of the above to start with. As your knowledge and skills improve in one of the areas, so you can take on more responsibility. However, you must never ever forget that keep a pig is a responsibility. You have a 'duty of care' to any animal you might keep no matter what the animal or what the end purpose for keeping that animal might be – a pig is no different in these terms from any other animal.

IT MUST NOT BE A ONE-SIDED RELATIONSHIP

Sadly, all too often, people take on the massive responsibility of becoming a small-scale pig keeper without looking any further than the end of their nose. In your case it is already marginally different because the very act of you buying this book demonstrates that you are 'pig serious' and not just 'pig silly'. The question you need to ask yourself first is not 'What pig is best for me?' but more importantly 'Am I the right person to keep a pig?'. Once you have accepted that the pig takes priority then you are taking the first tentative steps along the road to becoming a successful and, above all, happy pig keeper. Whether you are breeding, fattening or just keeping a pig as a pet you inevitably fail if you put yourself first. Remember that a 'pig treats you as an equal' so in order to get the best from the relationship you need to reciprocate fully.

Insight: The 'cute' factor isn't forever

I have spent more time in pig rescue centres because people forget pigs are not always small playful piglets. Puppies are not just for Christmas, and neither are piglets.

Basic requirements

Here are a number of questions you should ask yourself before committing yourself to being a pig keeper:

▶ *Have I got the right environment for a pig? Your top floor luxury flat with a magnificent view over Regent's Park may well be the ideal abode for you but it is totally unsuitable for any pig, even a specialist 'pet' pig. It is neither right nor sensible to keep a pig in the house with you. Despite the fact that they are easily trained, highly intelligent and very human friendly, pigs should not be kept permanently in your house. Not only is it unnatural but it is also dangerous because pigs are very hierarchical and may very well try to establish themselves higher up the household pecking order than you would want.*

Insight: Getting land

When I was looking for land I found that everyone wanted that nice, flat, well-grassed field but nobody wanted the overgrown roadside copse or small wooded area that was much cheaper and absolutely perfect for my pigs – they loved it.

▶ *Is my partner as keen on keeping pigs as I am? If the answer is 'no', 'I don't know' or 'I think so' then you should think again or re-address the question. It is no good assuming that you will never be ill, that you will never be incapacitated or you will never be placed in a position where you will not be able to look after your pigs. Even if it is only for a day or two, it is important from the pig's point of view that there is more than one person in the household who can step into your shoes, and do so willingly.*

Insight: Think of the bad weather as well

It is easy to look at your land in the spring/summer but I was soon aware that it becomes a different place in the rain and snow of the winter.

▶ *Do pigs take holidays? The answer to this question is obvious – no. Therefore you have to consider whether you are going to join the pig in its 'no holiday' regime yourself or, if not, what provision you are going to make for your pig in your absence. If you are planning on keeping a 'pet' pig and/or on breeding then it is imperative you understand that your commitment to the pig is 365 days a year. Even in your absence the pig has to be cared for, fed and looked after and unlike a dog, a cat or similar animal you cannot take it with you. It is not necessarily difficult to make such provisions but it becomes very difficult if you leave such planning to the last minute. If you are planning to act as a 'hobby farmer' or part-time smallholder, so that you and your family can experience the 'good life' with a better quality of meat in your freezer, then the question of holidays is less of a stumbling block because you can plan your new pig keeping venture around your holidays. Meat in the freezer does not require any looking after.*

▶ *Am I able to be patient and understanding? If the answer is 'no' then simply forget about being a pig keeper because pigs cannot be bullied and they certainly react badly to impatience. A pig is an extremely strong animal and if you try to push a pig or force it to do anything it doesn't want to do then it will always react by using its strength as well as its cunning to out-manoeuvre you. When it comes to a battle of wills then accept that your pig will always win if you show any signs of impatience. If you nurture any desire to use brute force then your pig will stop treating you as an equal and will instead highlight your inadequacies.*

Insight: Let other people make mistakes for you

Before I started keeping pigs I read a lot of books, spoke to a lot of pig keepers and used the knowledge I gained to make the mistakes I made less dramatic. I found learning from other people's mistakes a great bonus.

While these are some of the questions you need to ask yourself before spending any money there are a number of other questions you will want the answers to, and you may not know who to ask or where to look. So let us take a look at some of the more obvious ones:

▶ *Are pigs nasty? This is a question I am asked time and again by people who are considering travelling down the porcine road. Unfortunately, there is no simple response because in truth the answer is – they can be. A pig's temperament depends on a number of factors but primarily on you, the pig keeper. A pig is very rarely born nasty so in the great majority of cases if you see a nasty pig you must look firstly to the right of the pig, then to the left of the pig and then on either the right or the left you will find a nasty pig keeper! The two go hand in hand. Pigs are made nasty, but provided you follow the very simple rules set out in this book you need have no fears about your pig causing you any problems whatsoever. Treated properly they are easy to handle, easy to control and easy to train.*

▶ *Are pigs easy to keep and do they take a lot of looking after? Unlike a lot of animals, pigs do not take up a lot of your time or demand continued attention. As long as you practise good animal husbandry and your pig has a good, warm, dry ark to sleep in, your commitment will be limited to feeding twice a day with lots of TLC.*

Insight: Give yourself a breathing space

I bought my first three pigs before I had their pig ark complete thinking I could finish building it when they were in their pen. The result was a disaster and nearly a divorce!

Pigs come in a variety of shapes, colours and sizes but the one thing you will quickly learn is that they are highly intelligent and will respond to your kindness and attention more than almost any other animal.

PIGS AREN'T DAFT

Some are spotted black and white
Others black as the dead of night
Let no one even start to think
That a pig is only coloured pink

Some with snouts all long and pointed
Some squashed in and so disjointed
Some with ears that hide their eyes
Whilst others' ears reach to the skies

Large Black, Tamworth, Middle White
Make a truly awesome sight
All so different but don't ever
Forget the pig is very clever

Smarter, sharper, much more giving
Than any other thing that's living
Don't doubt my word – hear what I say
And stand and watch them for a day

It makes no matter what the weather
They all live happily together
British Lop and Saddleback
They have everything we lack

In the orchard two pigs grapple
For they've found a fallen apple
Gloucester wins and eats her prize
But Oxford knows it's undersize

Head down, tail up she makes a snuffle
Like she may have found a truffle
But no instead to her delight
A bigger apple that's just right

They root and dig in different ways
Turning minutes into days
Then lie contented in the soil
No more rooting no more toil

Grunting, talking sometimes barking
Piglets playing, piglets larking
Running madly round a tree
Just to celebrate they're free

They are so happy – don't wonder why
Because they're free not in a sty
Not for our pigs the chain or crate
They're free and really think it's great

A pig has a very knowing mind
And understands if treated kind
Will give you joy until the end
And treat you like a long lost friend

They know the time they know the sound
Of human feet upon the ground
Up jumps the Berkshire she's first off
In the race towards the trough

Who knows us better than the pigs?
One that watches, one that digs
They share our land and learn our ways
Whilst mother wallows piglet plays

They look at us through piggy eyes
Very thoughtful, very wise
The human they must truly think
Also isn't only pink

They view us kindly somewhat sadly
Because they see us made so badly
Noses so small they don't stick out
Like a proper rooting snout

Tiny ears they can't think why
They don't reach out up to the sky
And as for feet can it be true
All humans only have just two

They watch us clean they watch us feed
They watch us fill their every need
We know – to a background of piggy laughter
They say 'Is it us or them that's dafter?'

Tony York, 2002

The above poem should be a reminder to you that when you keep pigs it is a two-way occupation because you are being watched by your pigs as much as you are watching them.

10 THINGS TO REMEMBER

1 *If your partner and family are not in favour then don't do it.*

2 *Don't keep pigs just to 'keep up with the neighbours'.*

3 *Learn as much as possible about keeping pigs BEFORE you start doing anything.*

4 *Seek expert advice at the beginning and save money in the long run.*

5 *Making mistakes is costly so be thorough in your research and listen to those who know NOT those who think they know.*

6 *Decide what you are going to do, how you are going to do it and be sure that it is right for you BEFORE going to see piglets because they are addictive and might cause you to start keeping pigs for the wrong reasons.*

7 *A pet pig can live for 10 to 15 years of age and sometimes longer.*

8 Thinking *that it is right for you and your family is not good enough – you must be certain.*

9 *Never ask the person desperate to sell his/her pigs if they think you are doing the right thing ... of course they will say 'yes'. Discuss your plans and circumstances with an expert when you are thinking about it and BEFORE you make any definite plans at all.*

10 *Check your teeth because if you do decide to keep pigs you will certainly be doing a lot of smiling.*

2

Preparation and knowing the rules

In this chapter you will learn:
- *basic first steps that will set you on your way*
- *what you need to acquire before you can have your first pig*
- *the rules and regulations you need to adhere to.*

Basic first steps

Although the practical set-up procedure is covered fully in the next chapter, there are basic formalities you have to go through before you move on to the more practical side of becoming a pig keeper.

Firstly, if you do not have one already, you must get a County Parish Holding number (CPH). To obtain this number you have to apply to the Rural Payments Agency (RPA), informing them that you are planning to keep livestock on a small scale. They will send you the appropriate form to complete and you should receive your CPH number within about seven working days of you returning the completed form. Once you have your CPH number you are in a position to move forward.

Your next step should be to contact your local Trading Standards Office (Animal Health) and inform them of your plans. They will issue you with a Movement book (in which to enter details of any pig entering or leaving your premises) and they will also send you further very valuable information (dos and don'ts) about keeping your first pig. At a later stage they will visit you to see where you are keeping your pigs and offer further help.

Insight: Her Majesty's Prisons

I've always remembered being told to remember 'Her Majesty's Prisons' when I first asked what I needed to keep pigs. Looking blank I asked why. HMP was the reply: Holding number; Movement licence; and then you can have Pigs.

Finally, you can now start your research into what breed of pig will best suit your needs and where to obtain the help and advice that it is so important to get. You have already taken the first steps by buying this book but if you can get some practical experience as well then that would be an added bonus. Try searching on the internet for local breeders who you might be able to visit and talk to. The more knowledge and experience you can get before your first pig arrives on your land then the better for both you and your pig. Reading books and magazines about pigs and pig keeping is one thing but there is no substitute for some early hands-on experience. Just being among pigs alongside an experienced pig keeper will give you a whole new perspective and you will soon realize what amazing animals pigs really are.

Insight: Farm sales – my greatest days out

What on earth is a 'Farm Sale' I thought, because although I had seen a sign for one with lots of cars parked in a field I had never been to one. Take my advice: they are great events to go to and if you can avoid the bar you will meet lots of great people and often pick up some good bargains. But start bidding before you have been to the bar area!

Can anyone keep a pig?

Provided you have a suitable piece of land to accommodate the number of pigs you intend to keep, good housing and your pig/s cannot be considered or become a nuisance to those around you then the broad answer to this question is 'yes'. However, as a beginner, you need to know much more before you get your first pigs.

Government or EU rules and regulations about what you can and cannot feed your pigs, movement restrictions and requirements for identifying your pig change at regular intervals. You must therefore refer to the DEFRA website (www.defra.gov.uk) at regular intervals in order to keep up to date with any changes made. Do not worry if you do not have access to the internet because once you are issued with a CPH number your local Trading Standards (Animal Health Department) along with DEFRA will keep you regularly informed by post. DEFRA produce an excellent leaflet, 'A guide for new pig keepers', that gives superb advice and help as far as the legislation associated with keeping a pig is concerned. Ask your local DEFRA office or the Animal Health section of your local Trading Standards department to send you a copy.

Pigs and other animals

It is widely assumed that pigs and other animals do not mix well but in the large majority of cases nothing could be further from the truth.

Insight: Other livestock – friend or foe

A dear bank manager friend of mine who was fattening some Aylesbury ducks for Christmas soon found that if he slept pigs in the same building he would have no ducks for Christmas. Pigs are carnivores! Pigs and other four-legged animals (including horses, sheep, goats, cattle, dogs and cats) get on very well.

Although horses may initially find both the smell and the appearance of pigs (not to mention the strange noises they make) disconcerting, they eventually settle down. Horses are naturally flighty and will often rear up at the sight of a paper bag but once acclimatized to your new additions they get on very well. I know of many people who run their horse in the same field as their pig and in one case I know of a pig that acts as a 'guide pig' for an almost totally blind donkey.

Dogs soon realize that barking at a pig is a complete and utter waste of time because pigs treat such strange behaviour with contempt. Pigs and dogs of all sizes make great companions and cats, although naturally aloof, find pigs fascinating because despite a pig's size it makes no attempt to impose itself on either cat or kitten.

Chickens, ducks, geese and all wildfowl can run happily with your pigs in a natural outside environment where there is plenty of space and no restrictions. Very often you may see a chicken happily sitting on the back of a pig but don't be fooled because while in the open during the day everything is fine you must not confine them together in a building, especially at night. Pigs are carnivores and a sleeping chicken or duck makes a tasty meal. Beware also of running young ducklings in the same paddock as pigs for there are many stories of pigs picking off one duckling at a time as they dutifully waddle behind mum.

How much land do you need?

Although pigs do not need a great deal of room it is essential that there is enough space so they get sufficient exercise to keep themselves healthy both mentally and physically. They are instinctively very clean animals and will rarely dirty their own beds provided they are given an area of land large enough so they can designate part as a toilet area. You must remember that all pigs dig and root. It is a natural thing for them to do and although some breeds dig less than others you should never expect even the most docile of grass-eating pet pigs to act as a lawn mower in your orchard. They will certainly keep your grass down but more in the fashion of the Battle of the Somme than a croquet lawn. If you want to put your pigs in an area of a field, pasture or garden that you do not want rotovated then you will have to put a ring in their noses.

Insight: How much land do I need?

I soon learned it is better to have pigs in small patches rather than a big field. Pigs don't need a lot of land but moving them around is good. In my early days a 'pig sty' was plenty big enough. Moving pigs helps stop the land getting 'pig sick'.

Pigs are inquisitive animals and natural browsers so an area of rough overgrown land that they can root and dig to their hearts' content would suit them admirably. Do not make the mistake of spending hours, even days, before your pigs arrive clearing out any weeds, nettles or brambles – why waste your time and energy? Your pigs will relish the task and also do a much better job than you ever could. They are an ideal addition to any proposed new vegetable patch or to any area where vegetables have been in the past. A pig will not only turn over the ground for you and clear away many of the weeds but at the same time they will distribute a reasonable covering of manure resulting in a contented pig, a happy gardener and great vegetables. As descendants of the wild boar, pigs are natural woodland animals and therefore if you have an area of woodland, no matter how small, that you can use or incorporate into the area you designate to your pigs you will be creating a 'piggy heaven'. In larger woodland areas pigs are being used in increasing numbers to clear the undergrowth and brambles to help regeneration.

If you do not have an ideal piece of rough land or woodland that you can allocate to your pigs this need not stop you from becoming a pig keeper. In the early 1900s the 'back yard' pig, that was so important to the cottager, often only had a small 10 ft by 6 ft (3 m by 1.8 m) sty to live in so space should not be a problem. But even 100 years ago people realized the importance of giving their pig as much exercise as possible and at the same time allowing it to graze and browse. This was the time when the pig harness (now very rarely used) came into its own as a means of allowing the pig keeper to exercise and graze his pig on the grass verges which they affectionately called the 'long meadow'. Although this practice of harnessing a pig and then tethering it to an iron stake is now in the past it still clearly demonstrates that the amount of land required to keep a pig successfully is comparatively small. A good, dry, warm house and an area on which to graze and root is all a pig needs so long as it has your care, attention and commitment.

Obviously the precise amount of land you need depends on the number of pigs you have, how long you intend to keep them and their age. Remember that the longer the length of time you keep a pig or pigs then the more important it becomes to allow the land a period of time for resting. It is better to have three or four smaller areas and keep rotating your pigs from one area to another than it is to allow them access to a much larger area for a longer period of time. Allowing the land to rest will help prevent it from getting 'pig sick' and harbouring the potential to spread disease. A large garden is fine for a pet pig but if you are thinking of breeding then you will need a much larger area. If you want to fatten two or three weaners for pork once or possibly twice a year then an area the size of a tennis court would be sufficient as they are only going to be with you for about 18 weeks. You don't have to own a substantial amount of land to keep a pig but if you have rough, overgrown land or an area of woodland then you and a pig will make a perfect match.

Your first telephone call

As previously mentioned, before you can own and keep your first pig you must obtain a County Parish Holding (CPH) number which you obtain by contacting the Rural Payments Agency (RPA). Although you will make your initial request by telephone, informing the agency that you are going to keep pigs, you will not be issued with your own CPH number until you have completed the appropriate form that should arrive with you within days of your first initial contact. It is important to bear this in mind because taking into account possible postal delays, a two-way exchange of correspondence plus the time taken to complete your application it may be two weeks before you are allocated your

CPH number. Be well prepared and always apply for your CPH number before you select your pigs if you want to avoid being in a position where you cannot collect your pigs because you have not received the appropriate documentation from the RPA. The person you are buying your pigs from will not thank you if you are unable to collect your pigs on the day arranged because you don't have all the necessary paperwork. Because of the '21-day standstill' rule (see page 24) regarding the movement of stock on and off premises where livestock is kept any failure on your part to provide a CPH number could easily mean you have to wait a further three weeks before you can collect your pigs. You will not be happy at having your plans disrupted by getting much larger pigs three weeks older than planned while the seller is unlikely to be happy at having to keep and feed your pigs for longer than expected. The seller would be well within their rights to expect you to pay for 'feed and keep' during this period so don't let the situation arise – get your CPH number well in advance. It is a lifetime number associated with the property and land you are using and will never run out, but if you move properties you need to apply for a new CPH number.

The CPH number you receive will consist of nine numbers and they ensure traceability of pigs, for example if ever there is an outbreak of disease. It is an exclusive number referring specifically to your holding, farm, land or property. The first two numbers that are separated by a forward slash refer to the county where you are keeping your pigs. The following three numbers, again separated by a forward slash, refer to the parish within that county, while the final four numbers refer specifically to the land where you keep your pigs. No two CPH numbers are the same.

Your own herd number

Once you have received your CPH number your local Trading Standards Office (Animal Health) will be automatically informed of the arrival of your first pigs when you submit the white copy of your AML (Animal Movement Licence) but you need to register your pigs with DEFRA by contacting your AHDO (Animal Health Divisional Office). You can register your pigs over the telephone and provided that you give them your CPH number as a means of identification they will issue you with your own herd number either over the telephone or by post, usually within seven days. This number is made up of one or two letters followed by four numbers and is prefixed by the letters UK to meet with EU legislation. This herd number identifies any pig that is tattooed or tagged with this number directly to you as a further means of stock traceability. While a pig is not always required to be marked with this herd number for movement from one property to another, a pig must be marked with this number if going to the abattoir, irrespective of age.

Marking for movement

Any pig moving to of from your property must have an identifying mark of one sort or another. The first rule you should know is that all pigs, irrespective of age, must be marked with your herd number whenever they are going to the abattoir. This can be either with a metal ear tag stamped with your individual herd number or with a 'slap mark' (tattoo of your herd number applied to each shoulder of your pig with a 'slap marker'), both of which can be obtained from your local agricultural merchants. If you decide to use a metal ear tag then your herd number must be prefixed by the letters UK, but if you are using a tattoo on either the ear or shoulders the prefix UK is not required. A pig under 12 months of age moving to or from your property, provided it is not going to an abattoir, need not be marked for movement in this manner and is not required to be tagged or 'slap marked'. However, for movement purposes to and from any premises, other than an abattoir, all pigs under 12 months of age must have an identifying mark of some sort. A coloured cross, stripe or spot on the back or shoulder with a stock marking spray or paint as a temporary mark is sufficient provided this is indicated on your movement licence. All pigs over 12 months of age, irrespective of where they are being moved to or where they are located, must be clearly marked with your herd number using a metal ear tag, plastic ear tag or tattoo.

Movement licences and Movement and Medicine books

Before you can take your pigs home you must have an official Animal Movement licence (AML). This will be given to you by whoever sells you your pigs. You must not move a pig without a Movement licence because knowing exactly where all the pigs in this country are kept is extremely important. The traceability of pigs in the case of any serious outbreak of disease such as foot-and-mouth is vitally important as a means of stopping any spread of the disease.

Insight: Collecting and moving pigs

I have met some pig keepers who say they don't bother with all the paperwork. Don't fall into that trap because there is a very good reason for it. Simply protect everyone by making sure you have a 'Movement licence' when you collect or move pigs, and that all your pigs over 12 months old or going to the abattoir have your 'Herd number' on them. Your 'Movement book' must be completed every time a pig goes on or off your land. Traceability is important and saves pigs' lives.

The Movement licence is a four-part document with the bottom copy (yellow) retained by the seller and the remaining three copies are given to you when you collect the pigs. Within three days of getting your pigs home you must post the top copy (white) to your local Trading Standards Office (Animal Health) and enter the arrival of your pigs in an animal Movement book. The two remaining copies (pink and blue) of the Movement licence are retained by you provided that you collected the pigs yourself – if you used someone else to collect your pigs they keep the blue copy. You must enter the movement of all pigs on and off your land in a Movement book as soon as any movement of pigs takes place and the pink copy of the AML must be kept by you for at least six months. Although you will not require a Movement licence until you collect your first pigs it again makes good common sense to prepare in advance. Contact your local Trading Standards Office (Animal Health) and inform them of your plans and they will supply you with an official Movement book and will also probably arrange an early visit to explain all the procedures you need to know. They will also issue you with your own Movement licences in case you need to move your pigs from your land either to the abattoir or as a result of any sale you might make.

If you or your vet administer any medicines or drugs to your pig (including worming) you are required to keep a record of this in a Medicine book. Very often space to record such treatments is included in the Movement book but if not then an ordinary exercise book recording the date, identifying features of the pig, drug or medicine administered, quantity of dose given, any batch number of the drug used, how many days' treatment is required and the 'withdrawal time' will more than suffice. Periodically (possibly annually) you will receive a visit from an Animal Health Officer from your local Trading Standards department. They will usually telephone or write to inform you of their visit in order to ensure that it is convenient and you are going to be available on that day. They will want to check your Movement book and Medicine book to make sure that everything is in order and will bring you up to date with any changes in animal welfare legislation that might apply to you that you might be unaware of. Provided that you care for your pigs in a good and proper manner by practising sensible animal husbandry you should find your local Trading Standards (Animal Health) officials a source of great support, help and advice.

Insight: Medicine book

In my younger days when they used to feed pigs fish meal you could often taste fish in the pork. It stayed in the pig's body long after it was eaten and the same applies to drugs. Enter any medicines or drugs you administer to your pig in your Medicine book. Don't allow the food chain to become infected with animal drugs by making sure that you adhere to the 'withdrawal time' before that animal enters the food chain.

Special licence for pet pigs

Many people who keep 'pet' pigs are under the impression that because they are not a farm, smallholding or commercial enterprise they are exempt from all the movement rules and regulations but this is not so. All the Movement licence restrictions that apply to other pigs also apply to 'pet' pigs plus one more – a pet pig walking licence.

....
Insight: Walking pet pigs on a lead
Many people ask if they can walk their pig on a lead and the answer is YES but only if you get a 'walking licence' from your local Trading Standards Office (Animal Health) first.
....

Many people who keep small 'pet' pigs such as the Kune-Kune, Potbellied Pig or German Micro Pig want to walk their pig on a harness but this cannot be done without first obtaining a walking licence from your local AHDO. You will need to submit a detailed route that you intend to follow and submit this for approval. Your request may be refused if the plan you submit goes anywhere near farms where pigs are kept, near to livestock markets or even if your route passes by burger bars, fish and chip shops or other food outlets where your pig might find discarded food that could cause disease if eaten.

If your walking licence is granted, you must adhere to the route submitted and must seek renewal of your licence every 12 months.

Waste food rule

It is illegal to feed catering waste of any sort, whether from your own kitchen or from a commercial kitchen, to your pigs whether they are 'pet' pigs or not. Fresh or waste fruit and vegetables either your own or acquired from your local greengrocer, market trader or retail outlet can be used provided there is no direct or indirect cross-contamination with any products originating from animals. You must remember that although you may feed your pigs fresh fruit and vegetables that are good for them, you cannot store them or take them into your kitchen either wrapped or unwrapped whether they have come from your garden or from a greengrocer. All fruit and vegetables must be stored in your feed store, a shed, garage, outside or in some other room in your house NOT in your kitchen. EU legislation could well cause changes to this 'waste food rule' so you should update yourself regularly via the DEFRA website or local office. I cannot stress strongly enough the importance of never feeding your pigs meat, meat products or any food that has been in contact with meat. It is not just illegal to use catering waste from kitchens that handle and prepare meat – the same rule applies to vegetarian kitchens as well. Any contaminated waste food spreads viruses so beware. Do not even feed mouldy vegetables to your pigs or you could find your pigs' ears become brittle to such a degree that they break off. This is as a result of Ergot toxins produced by the fungus Ergot (Claviceps purpurea).

> ### Insight: My 'no go' area
>
> I have made my kitchen a 'no go' area as far as anything I am going to feed the pigs is concerned. It is illegal to feed your pigs ANYTHING that has been in any kitchen, including yours.

'21-day standstill' movement rule

Once a pig has been delivered to any premises you are not allowed to move any pig from those premises to anywhere else other than to an abattoir within 21 days. You are permitted to have more pigs arrive but whenever this happens it is always 21 days after the arrival of the last pig before you can move any pigs to other premises except to the abattoir. It is therefore very important that when you are collecting your first pigs you clearly understand that adhering to the exact day of planned collection means a great deal to the seller for they cannot have any pigs moving on to their land until after you have collected yours. Even altering your collection date by just one or two days could cause major problems. You should never consider that this 21-day standstill movement rule is just bureaucratic nonsense because it is not. These 21 days of 'standstill' act as a firewall against the spread of disease by covering the incubation period and therefore slows down the spread of disease. It is not only pigs that the 'standstill' rule affects and if you have other stock such as cattle, sheep or goats the arrival of your pigs will place a six-day 'standstill' on all these other animals. If you have any of these animals entering your holding then they place a six-day 'standstill' on all your pigs. In order to keep fully up to date with any changes to livestock movement rules contact your local AHDO who will send you a copy of the 'Rules for livestock movements' booklet. If you have access to the internet then visit the DEFRA website (www.defra.gov.uk).

Insight: Standstill rule

During the 'Foot and mouth' outbreak in 2002 many people wondered why they could not move their pigs. Quite simply it was to stop the spread of the disease and that is why there is a 21-day standstill rule. No pig except those going for slaughter may move off your premises for 21 days after a pig has come onto your premises. Twenty-one days is the incubation period for the major diseases.

Marking for breed registration

All the above rules and regulations regarding identification and the marking of pigs refer to UK and European law regarding the movement and ownership of pigs. But for a large number of small-scale pig keepers there is more to pig identification than simply applying and adhering to government legislation. The vast majority of hobby farmers, smallholders and those of you who just want to keep a few pigs to breed or keep a pig as a pet yearn for something different. They do not want an ordinary, run-of-the-mill, 'pink pig' that is more used to a commercial farming environment because they are seeking much more than that. Being able to return to the older, more traditional breeds brings with it memories of the past that are undoubtedly very appealing to so many people and if you decide to take a similar route you may need to have additional markings on your pigs.

If you choose to keep or breed one of the recognized rare or traditional pig breeds then not only will you be fulfilling your pig keeping dream but you will also be playing a significant part in helping save these wonderful breeds from extinction. Just because these breeds do not necessarily meet the requirements of the larger commercial pig units where the demands of the supermarkets take priority that is no reason why they should become obsolete. If you intend to keep and breed pigs that fall into these categories then you must join the British Pig Association (BPA) or one of the other two bodies (British Lop Society and the Kune-Kune Society) that handle and control registrations and birth notifications for these breeds. Pure bred and registered pigs, no matter what the breed, have to be marked and clearly identifiable as registered and pure bred. When you join one of these controlling bodies you will select your own herd name (traditionally linked to your farm, smallholding or property) and provided this name is not already in use you will then receive your own 'herd designation letters'. These will then be used on all official registration certificates and when marking pigs for breed identification. Pure bred registered pigs must be ear tagged, ear notched or ear tattooed according to each individual breed society's requirements. You will receive all the help and advice you require on this matter from each individual breed society.

Insight: Where should I go?

To start with I relied on farming neighbours to point me in the right direction to buy tattoo equipment and other necessary things. I had never been in an agricultural merchants in my life before but once introduced it was difficult to get me out! I was like a kid in a toy shop!

Breed societies and clubs

All the recognized rare and traditional pig breeds have their own club or society that you can join if you wish to. They all have an annual general meeting, usually have a regular newsletter and sometimes have a Christmas luncheon and maybe even organize trips to agricultural shows. They do a great deal to support and promote specific breeds and although joining such a club or society is not essential (with the exception of the British Lop and Kune Kunc societies if you are breeding pure bred registered stock) you will gain a lot of support and help as far as your particular favourite breed is concerned if you do join. If you live in Wales or in any of the English counties bordering Wales you will find the highly effective Wales and Border Counties Pig Breeders' Association is always there to help and support you no matter what pig breed you decide to keep. You will find the association is prominent at such shows as the Royal Welsh and the Smallholders' Show both held on the Royal Welsh showground at Builth Wells in Powys.

Moving a pig anywhere

There are certain things you should remember and understand when moving your pig or pigs whether collecting or delivering:

▶ *Moving a pig is about cleanliness and good hygiene. Stopping the spread of disease or infection is paramount so make sure that your vehicle is spotlessly clean both before and after you have transported your pig/s.*
▶ *Moving a pig is about consideration and caring for your pig/s. Make sure that you keep this uppermost in your mind when you load your pig. Whether large or small your pig/s need to be as comfortably secure as achievable in as small an area as possible. Do not give them so much space that they are thrown about on the journey in a manner that could cause sickness.*

- ▶ *Make sure your Movement licence is completed correctly and the pig is identified either by a coloured mark (under 12 months of age) or a metal tag or 'slap mark' if over 12 months old or, if going to the abattoir, at any age.*
- ▶ *Make sure that you are not restricted moving a pig OFF your premises by the 21-day standstill rule. No pig may leave your premises until after 21 days have elapsed since you had a pig enter your premises. This does not apply if you are taking a pig to the abattoir.*
- ▶ *Enter all movements to and from your premises in your Movement book.*
- ▶ *Send the white copy of your Movement licence to your local Trading Standards Office (Animal Health) within three days of a pig arriving on your land.*
- ▶ *An adult pig that is too big to lift comfortably and put in a dog cage or similar MUST travel in a proper stock trailer that has gates which close before the back ramp is put up. Keep it as confined as possible.*
- ▶ *If you have a young piglet you can comfortably pick up, it can travel in a dog cage or similar that can be cleaned and disinfected before and after the journey. Keeping the piglet confined so it does not get thrown about on the journey is good practice.*

10 THINGS TO REMEMBER

1 You cannot keep pigs until you get a CPH (County Parish
   Holding) number from the Rural Payments Agency in
   Newcastle. You must have this BEFORE you arrange to
   collect your pig/s.

2 You must get a Movement licence from whoever you get
   the pig from when you collect your pig.

3 Not all pigs are the same. Make sure you get the breed that
   will suit your needs.

4 You MUST send the top copy of your Movement licence to
   your local Trading Standards Office (Animal Health) within
   three days of collecting your pig/s.

5 Within a week of sending in your Movement licence or
   informing the AHDO they should send you an animal
   Movement book and a Medicine book plus a herd number for
   your pigs. If not (probably because they are very busy) give
   them a call so you get what you need to stay within the law.

6 Always keep to your agreed date for collection with the person
   you are getting your pig from. Any change may affect his/her
   plans as far as the '21-day standstill' rule is concerned.

7 Make sure the land you are using for your pigs is clear of
   glass, barbed wire, tin cans, bottles or anything that could
   cause damage to your pigs. Land that is overgrown with
   nettles, weeds and brambles is fine but not discarded rubbish
   that could harm them.

8 NEVER feed pigs anything that has been in your kitchen.
   If you get fruit or vegetables to feed your pigs (very good for
   them) keep it in your feed store, conservatory, dining room,
   outside shed or even your lavatory but not in the kitchen.
   That is the law.

(Contd)

9　Always enter all pigs that come onto your land or that leave your land in your Movement book. It is easy and simple to do and your local Trading Standards officer will come to check this book every so often, so keep it up to date.

10　You cannot walk a pig on a lead without first getting a 'Pet pig walking licence' from your local AHDO.

3

Before your pig arrives

In this chapter you will learn:
- *how best to house and contain your pigs*
- *the basic things you need to provide for your pig*
- *about rooting and digging – a pig's natural way of living.*

Having a suitable piece of land and knowing that you are the right type of person to take on the responsibilities of a pig keeper moves you on to the next question – what now? Before your pig arrives there are some things that you must have, some things you might have already, some things you should have and some things you need to think about for the future.

MUST HAVE

▶ *A pig ark (house) to sleep in.*
▶ *A feeding trough.*
▶ *A water bowl or drinker.*
▶ *County Parish Holding Number.*
▶ *A means of getting your piglets home.*
▶ *Suitable bedding such as straw or hay.*
▶ *Good secure fencing.*

MIGHT HAVE ALREADY

▶ *Buckets (for carrying feed in).*
▶ *Fork and spade (for mucking out).*

- *Wheelbarrow.*
- *Overalls, wellington boots and practical working clothes.*

SHOULD HAVE

- *Board for guiding your pigs.*
- *Coloured marker spray (for identification when needed).*
- *Thermometer (preferably two).*
- *Wormer (to worm on arrival if necessary).*
- *Registration with a local veterinary surgeon.*
- *Suitable insurance cover.*

Insight: New arrivals

Oh! It is so easy to get piglets home but not so easy to keep them there. I have watched many, many new pig keepers fall into the 'Now we are home' trap and fail to remember that while *they* are home the pigs are somewhere completely new. They will be in a strange place, bewildered and disorientated by the drive so make sure that wherever you put them for the first night is really secure. My first pigs escaped within 15 minutes of arrival.

Fencing

Having selected the area where your pig is going to live, it is no good whatsoever providing ideal sleeping accommodation if you have failed to recognize that the fencing surrounding their home is less than secure. You can have the most perfect house, ideally located in the most select position possible but it is of no use at all if your pig spends all of its time digging up other people's land or neighbouring gardens. Your first priority must be good, secure fencing.

If you think that your pig will stay happily where you put it with just the minimum of fencing – you are wrong! Your fencing needs to be not just stock proof but also 'pig proof'. If you have to keep your pig inside while you improve your fencing then so be it but whatever you do don't expect your pig to turn a blind eye to any highly inviting weaknesses in your fencing system.

Insight: Escape from Alcatraz

Many times I have watched newly arrived pigs walk the fence looking for a weak spot and checking the gate to see if they can lift it off the hinges. If there is a way out they will find it. It is wise to keep them shut in their ark for the first night or in a very secure place until they get used to their new home. Barbed wire at the bottom of the fencing and gate hooks reversed always helped me.

Good pig or sheep netting firmly erected should suffice provided you have a good strand of barbed wire (two strands are better!) at the bottom of the fence. This will stop your pig from rooting under your fence but make sure the barbed wire is on the inside of the fence not the outside. Naturally if you 'ring' your pig this will further deter its desire to root under the fencing and make 'pasture grazing' possible and less of a headache.

Try not to get into the habit of feeding your pig by the fence for this will also encourage your pig to pull at the fence as you are approaching. I am sure that you get pretty excited about your food so why should your pig be any different?

Electric fencing is a great asset when it comes to keeping pigs in one place. You can use either individual strands of electric wire or electric netting but if you use strands of wire then you must educate your pig to recognize what it does if it has never seen electric fencing before. The process of educating your pig is quite simple and takes only about an hour or so. Make sure that you start by putting your electric fence immediately in front of a secure stock fence or wall with the wire or netting being the brightest colour you can get so it is easily seen. The area you use for training or educating your pig need not necessarily be the area where it will eventually be housed and need not be a particularly large area because this is just a recognition and training exercise. For the purpose of educating your pig you need only have the electric fence along one side of a secure area. When your pig goes to examine the fencing (you can be certain that the first thing a pig will do in the first hour, when put in a new area, is to walk the perimeter of the fence searching for any weak spots) it will get a shock and jump back and after two of three such shocks will have learned to keep clear. If you don't educate your pig in this way by having your electric fence initially backed up with a stock fence or wall behind it then when your pig gets its first shock it will automatically jump through the fence! If you fancy an hour or two of chasing your pig and trying to get it back inside the electric fence again then ignore this advice, but do so at your peril!

Insight: The Electric Shock treatment

When I first kept pigs in a wooded area I decided it would be a good idea to 'strip graze' them, moving them to a new strip every week. In order to do this I decided to use electric fencing. What a disaster! The pigs did not know what it was so got a shock and immediately jumped through it!! It took me several hours to get them all back again. Make sure you follow the advice in this book and spend half an hour training them how to respect electric fencing before letting them loose with it.

Pigs are very respectful of the electric fence once they have inadvertently touched it a couple of times. They are intelligent enough to keep well clear in future. You may find it very difficult moving your pig across an area where you have previously had an electric fence even though you have taken it down. One tip is to put a path of straw down where you want your pig to cross and it will then walk where the wire has been with a little encouragement from a bucket of pig nuts, an apple or two or some other tasty morsel. Never try to drive or push a pig – you will lose and waste a great deal of time. Lose your patience with a pig and you lose everything.

You do not need a complicated electric fence system, just something simple that you can run off an old car battery.

Your local car dealer or garage will let you have an old battery with enough strength in it to operate your fence for nothing or at worst very little. Plastic wire, fencing posts, insulators, a battery and the energizer is all you need. Unless you want to go into the 'big time' there is no need for you to have a mains system, winders, reels, gate hooks and all the other paraphernalia that is available. If you have a simple system that is effective, easy to operate and easy to move – what more could you want?

The entrance to your pig area

No matter where you decide to keep your pig you will need an entrance gate of some sort. Whether this is just wide enough to get your pig in and out or whether it is also large enough to drive in with a trailer, you must remember one thing – reverse one of the gate hinges! Pigs have remarkably strong necks and noses making them world experts at putting their noses under even the largest gate and lifting it off its hinges to escape even at the youngest of ages. If they cannot find a weakness around the perimeter of the fence then they will automatically try the gate.

Insight: Good fencing is a priority

I found out right at the start that a professional fencing contractor saves rather than costs money in the long term.

Pig housing

I know that many pig keepers provide 'temporary' housing for their new pigs while they organize more permanent accommodation. It is easy and simple to house a young piglet but all piglets will soon grow much bigger and therefore we hope the following advice will help.

Most rare and minority breeds are bred to be kept outdoors and are therefore very hardy. They can cope with all weathers and will thrive under conditions that many intensively reared breeds would shy away from. Once you have seen a hardy, traditionally reared old breed of pig successfully farrowing in her straw-filled ark while it is snowing heavily outside you will understand what I mean. Young piglets happily playing in the snow is a sight to behold.

Having said that the traditional old pig breeds are hardy they will still always need a good shelter – somewhere dry and warm where they can go to sleep, take refuge from the wind and rain as well as getting protection from the sun during the summer months. Most importantly, it must be draught free because although your pig will deal admirably with the cold and wet, no pig likes draughts. Wind is no friend of a pig.

While pigs do not need five-star hotel accommodation even the very hardy outdoor breeds need good shelter from the elements. You have the choice between a permanent home which is static and stays in one place or temporary movable housing such as a traditional wooden pig ark. Permanent pig stys are usually made of concrete, bricks and stone blocks with a corrugated roof adjoining an exercise yard containing feeding trough and water. A modern movable ark is almost always timber framed with corrugated sheeting or timber boards for the roof and sides.

In most outdoor situations a simple ark of corrugated metal sheeting will do but the better the home you provide for your pig then the better will be the end result. A metal ark has the disadvantage that it is cold in winter and hot in summer (not an ideal situation!). Many metal arks do not have floors with the end result that during the winter they create wet, cold and muddy conditions for the pig and unnecessarily difficult cleaning out problems for the pig keeper. If you have a weatherproof shed or outbuilding you can soon turn it into an ideal home for your pig at very little cost at all. If you are lucky enough to have an old 'tailor-made' brick pigsty on your land then both you and your pig have a head start. But remember if you farrow in the pig sty make sure there is a covering of straw to protect the new-born piglets from dying of hypothermia on the brick or concrete floor during the first 24–48 hours. It is not unknown for newly born piglets that are wet, and still trying to master the art of walking more than two or three steps without falling down, to lay down to rest only to find they are virtually stuck to a bare, cold concrete or brick floor.

The thing that you must remember about accommodation for your pig is that what it looks like doesn't matter but how it performs does. We have seen many weird and wonderful contraptions converted into pig houses and the strangest parts of buildings accommodating the happiest of pigs. Provided a pig has room (and remember that the small piglet you bought will grow into a big pig) to move around, protection from the elements and a dry, well-bedded, floor then you will have a contented animal. That is of course provided there is exercise space outside and a good diet to accompany it. Space and size is not only important for your pig – it is also important for you.

There is nothing worse (I know from bitter experience) than being bent double, Quasimodo style, dust in your lungs and the smell of muck in your nostrils, trying to clean out a pig sty. At the same time your favourite pig will undoubtedly be giving you all her love and attention by putting her full, not inconsiderable, weight against you as a sign of affection. It is at times like this that you will wish you had taken note of our advice to allow plenty of room for your pig and headroom for yourself. A good pig house should be roomy yet cosy and dry in winter yet not too hot in summer. Allowing ventilation for air to circulate is extremely important but avoid draughts.

If you want to buy a ready-made pig ark then start looking early. If you don't leave it to the last minute then you may be able to find suitably priced secondhand arks advertised in your local paper or at a local farm sale. Many traditional farmers are turning their backs on pig breeding in favour of more financially rewarding subsidized activities and are pleased to be able to find a buyer. Anything between £150 and £550 should get you a very good ark depending on whether it is new or secondhand, the condition, how easy it is to move and how desperate the current owner is to see the back of it. If you are very lucky sometimes you will be offered an ark at a much cheaper price but sadly such bargains are now few and far between. If your new ark needs repair work to make it weatherproof then value it accordingly. The work putting it right will not be difficult and you might well end up with a real bargain. The one problem you may have is moving a large, sometimes cumbersome pig house from one place to another. The cost of transportation can often be as much as the ark itself so if you do not have your own trailer check first on the costs involved in getting your new acquisition home. Obviously if the ark is designed to be taken apart and then re-erected it is a very good investment as it will be easy for you to transport and you will also be able to move it around your land as required.

Whatever type of shelter you provide there are some basic requirements that you must provide. Not only must your pig have a waterproof home that is dry all the year round but I stress again **you must also ensure that there is protection from draughts**. Outdoor breeds are bred to withstand the cold but a cold, wet pig with no shelter from the wind is a decidedly unhappy pig.

If there is no door and little or no front to your pig ark it is a simple matter to hang a sack over the entrance for use in windy weather. Your pig will get a degree of protection yet be able to push it aside to get in and out. An old door picked up at a farm sale or your local sale room also makes a very useful baffle when stood lengthways 3 or 4 ft (0.9 or 1.2 m) in front of the entrance. Although your ark may have no door, a partial screening either side of the entrance will enable your pig to keep out of any through wind provided the ark is big enough.

Insight: Doors are important

Some people ask why I have doors on all my pig arks. One reason is to keep them confined and secure on their arrival but I often need to separate pigs for other reasons so I always make sure all my houses have doors that close securely.

As far as the floor is concerned, if you have your outside shelter on good dry land with plenty of straw you need have no floor and can set it down on the bare earth in the summer. Sadly, in this country we have our fair share of rain. If, therefore, your land is wet then it is imperative that you have a good wooden floor on any outside hut or shelter. It is quite easy using plywood and/or surplus boards to make a good floor in even the most basic of corrugated arks and thereby ensure a dry, warm base on which your pig can lie happily among the bedding.

If you are any sort of a 'handy person' then making your own pig ark will be both fun and cost effective. A wooden ark with an 'A' frame and a good wooden floor can often be made out of spare timber that you were wondering what to do with. There are some things you should always remember when building an ark of your own:

1 *Make it big enough inside to accommodate both you and your pig. When it comes to farrowing and cleaning out you will curse if you don't.*
2 *If your ark is smaller than is ideal then a large opening with a sliding, wooden shutter set at the back of the ark will allow you to observe your sow farrowing without disturbing her. If you make it low down and large enough you will also be able to reach in and help her if the situation demands.*
3 *A door on the front that can be shut and bolted is not just a useful addition but should really be an essential part of any ark. There are many reasons why you should incorporate a door as a standard addition on any ark. Firstly, it will enable you to confine your pig if the vet is needed or if you wish to examine your pig yourself. Secondly, you will be able to shut in the piglets when they are very young while mother has a break or feeds. Thirdly, you can close the door at night to protect very young and possibly vulnerable stock. Finally, it enables you to keep out any driving snow or rain when our unpredictable climate is at its worst.*
4 *A wooden ark standing on old railway sleepers or timbers to keep it off the ground protects the timbers from rotting while*

*also giving good air circulation. You will also be creating
a marvellous new play area for any young piglets as they
scamper underneath their house playing the piggy equivalent
of 'hide and seek' during balmy summer evenings.*

5 *You may want to move your ark from time to time. It is easier
to move an ark that is securely bolted together and it will also
last you much longer. Avoid using screws and nails because
pigs are strong animals even when they are comparatively
small. It is not unknown for a fully grown pig to demolish a
poorly constructed ark simply by rubbing against the side in
an attempt to ease some irritation or another.*

6 *If you use corrugated iron you must remember it is hot in
summer and cold in winter. Therefore it is not the best
accommodation unless you have good insulation in the winter
and good foraging land with shade in the summer.*

7 *If you have good dry land that drains well then an earth floor
for your outside ark will be satisfactory providing that it
doesn't get wet when it rains. Personally, I prefer not to leave
anything to chance and always opt for solid wooden floors
that are dry and give good insulation.*

8 *Have all the timbers pressure treated (tanalized) before
construction. This will give your ark many more years of life and
will also help prevent your pigs from chewing at the timbers.*

9 *Two doors. A door at the front and a door at the rear are
definitely two of the most important features you should
incorporate into your ark. This will enable you to use your
ark for larger pigs than would be the case if you just had one
entrance because they have no need to turn round to get out.
Pigs are notoriously bad at backing out of any area and with
both a front and rear entrance they can go in one way and out
of the other. When used for smaller pigs the two-door ark will
enable you to locate your ark so that it is midway between any
fencing (electric or otherwise) with the front end of your ark in
one grazing area and the rear end extending into the adjoining
area. You are now in the happy situation of not having to
move your ark or fencing when you want to move your pigs
to a fresh patch. Simply keep the back door on, with the
front door open, when you want them to graze the land at
the front of your ark and then reverse the situation when you*

want to give them access to the fresh ground at the rear of the ark. This is a very easy way of rotating your pigs and stopping any one area from being over used or becoming 'pig sick'.

If your pig has access to the land and is only using its house as shelter and to sleep in, your work mucking out will be kept to a minimum. Your pig will not dirty its bed and the straw will last for longer than any other animal you might keep. You have no need to clean out your ark by loading the used, broken straw and dust into a barrow before taking it to a suitable manure heap because there will be no muck. Simply take a long handle broom and sweep all the old straw out onto the ground in front of your ark. This will then act as a doormat for your pigs before they enter into their newly bedded house. A floor in an ark makes cleaning out so much easier for you.

No matter what type of pig ark you ultimately decide to give your pig, always remember that pigs are remarkably clean animals and will never dirty their bed. However, if you provide the wrong sized ark you will be encouraging them to dirty their bed and will be creating more work for yourself cleaning out their ark. NEVER buy an ark that is too big for your pig/s. If the ark is too big you will also waste your money.

If you provide an ark that is too big you may think that you are being kind to your pig/s by giving them 'luxury' accommodation but they will disagree and probably do the following:

▶ *All sleep in one corner cuddled up close together for company and warmth in a manner that means much of their ark is not used for sleeping in, thus wasting your money.*
▶ *They will turn the area into an 'en-suite' by sleeping in one area and then because they have such a large area they will create a separate toilet area rather than going outside to prevent dirtying their bed as would normally be the case.*

What should you do if you plan to start with small piglets (weaners) and move on to breeding or fattening at a later stage? Such progression would result in you having bigger pigs at a later stage that will not fit in the small ark. If you have this in mind you do not want to be faced with buying another ark so you should buy a larger ark that suits all stages of your planned pig-keeping venture. This will save you the money that you would have spent buying a larger ark later on.

'But what about them turning such a large area into an "en-suite"?' I hear you say. The answer is very straightforward. Simply put as many bales of straw in your ark as is necessary to fill the excess space leaving your pigs with just enough space for their bed. As they grow so you can remove a bale or two so they have sufficient sleeping space and then eventually when they are fully grown you will have an unsoiled ark free of straw and clean toilet-trained pigs. If you do not want to use straw bales then any other method of dividing up your ark as they grow will do.

ARK COMPARISONS

1 **Metal arks.** *These are comparatively cheap but can be extremely cold in the winter and boiling hot in the summer, which is not the ideal combination. They usually come without a floor, or floors are an 'extra' – the same applies to doors. If they are not well made and sturdy then a big pig will soon knock it about. A lack of flooring can be a big problem because the pig gets cold and wet in the winter, often digging out a 'pit' to lie in. This can fill with water or mud and is most unpleasant in the winter. More important, it makes your job of cleaning out very difficult. On the plus side, they are fairly easy to move with a tractor or several willing hands and are usually cheaper than most other types of pig housing. In life you generally get what you pay for and what you see is what you get. I think doors and floors are so important that they more than justify any extra cost.*

Insight: Gently cooked pig

The first ark I had was a metal one and as it was in the middle of the summer I nearly cooked my first pigs months before they were ready without even realizing it.

2 **Plastic arks.** *These are usually light and quite easily moved. Unfortunately, your pigs might find them as easily moved as you do. If your pig has outgrown the ark then beware. A big sow will toss it in the air unless it is well pegged down! Sadly, some cheaper plastic arks become brittle and crack when outside in the elements and therefore you need to make sure that the material used is truly going to last for more than two or three years. Good, well-made plastic arks are excellent so pay a little more and get the best you can afford because it will be worth it in the end. Often floors are not included and can be an additional cost or more on top of the price of the ark which can make them more expensive than they may at first appear.*

3 **Wooden arks.** *Wood is warm in the winter and cool in the summer, giving an almost ideal environment for the pig. Always make sure that the wood is treated (tanalized) before construction in order to increase the life expectancy. Every year treat your ark with a water-based wood preservative and it will last you many years. Make sure that any preservative you use is not harmful to animals and if possible keep your pigs in a different area while it is drying.*

Remember that you would not like to live in a badly constructed, insecure, draughty house with a muddy floor, no door and without adequate bedding so you shouldn't expect your pig to endure such conditions either. If the land where your pigs are being located is undulating, on a slope or without a suitably flat area on which to locate your pig housing then set your ark on timbers that have been set level in the ground.

A permanent building or an old tailor-made (only if you are a dwarf!) pigsty will often have a concrete floor and concrete run. Be aware that concrete is both cold and slippery. A pig sliding around on a wet, messy floor can soon damage itself. A wet, cold concrete floor without ample straw or other covering is a recipe for disaster.

One final word of warning. A pig (even a small one) can be a very strong animal and you should not overlook this fact when providing its home. A temporary 'lean-to' will soon become a permanent heap of rubbish once your pig starts rubbing against it or rooting under it. Even the most docile of pig breeds are curious and have an element of devilment that will test the sturdiness of its home as well as your overall fencing capabilities!

BEDDING

Dry clean bedding is important. A pig is not a dirty animal and will never soil its bedding provided there is a space outside the ark that can be used as a toilet area. Even in a traditional pigsty or confined area a pig will always use the outside run or one corner of the building specifically as an area for waste. If ever you see a pig with dirty bedding or lying in its own muck it is the pig keeper that is to blame, not the pig.

Good, clean straw changed regularly will provide warmth and protection as well as keep your pig clean and dry. Some pig keepers use sawdust and fine wood chippings very successfully as, although this is much more expensive than straw, it does absorb the moisture better. However, it can cause a disposal problem as it does not break down so easily when disposed of. Well-rotted dung and straw make excellent manure for the garden whereas sawdust or fine wood chippings do not. Over a period of time your pig will break up the straw as well as eat some of it and pig houses often become very dusty especially if outside on open ground. It is therefore good husbandry to periodically clear out the bedding completely and start again. You may find a face mask against the dust necessary and if you find this is required just think how pleased your pig will be once you have cleared out the dust.

Get in a few bales of straw in preparation for the arrival of your pigs and if you do not already have a ready supply make sure you know where to get good dry straw on a regular basis for the future. You will not need a lot but be prepared. An inside pig in a sty or comparatively confined area needs to be cleaned out on a daily basis while a truly outdoor pig that only uses the ark to sleep and rest in requires much less attention. Fresh straw as needed, dependent on both the weather and the time of year, plus a periodic spring clean will suffice for there will be no droppings or mess to clean out.

If your pigs are outside and running free then a little straw outside in front of the entrance will help to keep any excess mud and dirt out of their house. You have a mat to wipe your feet on so why shouldn't they?

Insight: Playing the straw game

Seeing how quickly they could take the straw and bedding from their pig ark after I had put it in became a game that they would have played all day. My frustration was slowly turning to anger until I realized that they stopped doing it when I moved away and stopped putting it back in. Obviously their game was no fun without me joining in!

Water

Pigs drink a lot of water and therefore you must not forget this when considering where to locate your pig. If you have to constantly carry two full buckets of water (which after a few yards begins to feel like lead!) any real distance at all from the water source to the trough it can soon take the gloss off pig keeping.

The perfect answer is to have water constantly piped into the pig area so that fresh water is always at hand, but it is not necessary. Provided you remember that the nearer you keep your pig to a water source the easier it will be for you then all will be well. If you are lucky enough to have a stream, river or spring in your pig patch then you are truly heaven blessed. If you are keeping a number of pigs then you might like to consider installing an automatic water system with a number of individual drinkers. Your pigs operate this system themselves by pushing a pressure pad with their snouts in order to release the water into a small drinking bowl. Pigs are very intelligent and quickly learn how to operate this system which, while expensive, does provide a constant supply of fresh water with no trough to clean out and fill twice a day. Whatever your circumstances there must never be a lack of water both for drinking and, if kept outside, for making a mud wallow during hot weather.

The pig is one of the only animals, other than the human, that suffers severely from sunburn and heatstroke. A mud wallow is an essential provision if you want a really happy, healthy outdoor pig but do not waste your time and energy digging one for them. Provided you make the water available in a particular part of the area your pig is using, your pig will do the rest. They have strong necks and snouts so are very well equipped for the job but much more importantly they will produce a much better end result than you would with a size and shape to suit their exact needs. If you find that your pig is regularly tipping over the water trough and trying to lie in it then you are being told in no uncertain manner to provide water on the ground as well as in the trough.

Insight: Making a wallow

Before I got my first pigs I often pondered on how and where I should put the wallow that was so necessary for them in the summer. I should never have worried because the minute they saw their wonderful new water trough they just tipped it up and made their own. They dug a better wallow than I ever could. In the summer put the water trough in the centre of the pen so that when they tip it up they don't weaken the fence line.

There appears to be a developing trend to use sun blockers on pigs in the hot weather but this is not really necessary. Simply ask yourself what a pig did before we came along with the idea of using such a very unsuitable and costly method of sun protection. Provided that you have allowed your pig to make a wallow and have provided the correct housing where your pig can get shade from the sun (don't forget that metal pig housing can get extremely hot under the strength of the summer sun) then you are unlikely to face any problems with your pig getting sunstroke. Having water close to where your pig is located along with a hose will enable you to spray your pigs in really hot weather. Not strictly necessary but a delightful interlude for both you and the pigs.

A little more thought

With a good feeding trough, a water container and stout, well-erected pig fencing surrounding the area where your pig will run you might think that you are ready to start but there are a couple of other factors to consider, regarding feeding your pigs.

Firstly, your water and feeding troughs need to be very heavy or set into concrete. Your pigs, even at an early age, will be both strong and mischievous. Pigs will soon upturn a light trough causing all sorts of chaos on top of the frustration you will feel at having to refill water troughs time and again. If you ever thought that feeding time at St Trinians was something to behold you have seen nothing until you have witnessed two pigs battling for the remains of an upturned feeding trough (especially if you are feeding wet swill).

The problem is not so marked if you feed concentrated pig pellets. If you are feeding outside on reasonably dry ground then you can do without a feeding trough. By scattering the feed on the ground among the grass, weeds and undergrowth you will not only be feeding your pigs but you will also be keeping them occupied for the next hour or so as they hunt for their food. If you have rough, weed-covered land that you want cleared then the process of throwing their pig pellets in this area will encourage them to clear it even more quickly for you. Your pigs will enjoy the process of 'food hunting' and anything you do that makes the day that bit more interesting for your pigs will result in your pigs being more contented. Pigs only develop bad habits or vices if they are bored and are restricted to a small area with overcrowding, which often results in tail biting.

When the ground is wet and muddy you can still avoid having to use light feed troughs that are easily turned over by putting half a dozen paving slabs down in the field or area where your pigs are. These slabs make an ideal feeding point and can be easily brushed clean or hosed down every day or so. Unlike troughs, pigs do not turn paving slabs over and lose all their feed in the mud. Remember rain plus pigs equals mud!

Good old cast iron troughs can often be obtained from farm sales but unfortunately the antique dealers are pushing the prices through the roof. They sell them on as plant troughs and garden ornaments – but keep your eyes open and you might be lucky. These old iron troughs have the added advantage that they have separate eating areas divided by metal bars thus preventing one pig from bullying another and getting more than its fair share of food. Being solid cast iron they are extremely heavy which prevents your pigs from easily turning them over but the disadvantage is that they are also extremely heavy for you to move any distance – not a job for the weak or weary! Another alternative is the old white, very heavy, Belfast kitchen sink. One can often be bought from your scrap merchant, builders' yard or found at a farm sale. If they are in perfect condition they will not be cheap but if you find one that is slightly damaged or badly chipped it will still suit perfectly as a feed trough. A little cement in the plug hole quickly makes it perfectly serviceable as a water trough. A bucket set in cement or half a land drainage pipe also set in cement also serve the same purpose.

For younger pigs (if you start with weaners) a useful tip is to use the top of an old milk churn. Fill the centre with a large stone or stones and put the water or feed around the rim to make a very cheap, sturdy and serviceable feeding trough for young piglets. Nailing the upturned tops of old sweet jars to a stout piece of wood or old railway sleeper serves a similar purpose if you have a number of young ones to feed. An excellent 'feeding trough' for young piglets when feeding their creep feed in a secure section of the farrowing pen is a simple house brick – it is too heavy to turn over and has an excellent hollow on top in which to place a little feed.

Insight: Food and drink galore

The one thing I hadn't allowed for when I got my first piglets was the amount of food and drink I would need. NOT FOR THE PIGS but for all our friends, neighbours, relatives, work colleagues and 'uncle Tom Cobley and all' who came to see the new arrivals that first weekend. The cost of supplying food and drink for my two-legged friends far surpassed what was spent on the pigs but I soon learned that this was a perfect time to take orders for the 'mouth-watering' pork and 'crackling that crackles like it used to' that would be available in 17 weeks time.

Digging

Before you collect your pig or pigs you will need to fully understand that digging and rooting the ground is a very natural thing for a pig to do.

Although I know of a number of pig keepers who still put a ring in the nose of their pigs to stop them digging and rooting, the practice appears to be gradually dying out. From an animal welfare point of view, ringing the nose of a pig has nothing to commend it at all. There can be no argument that it causes the pig pain because if it did not then it would not stop it from digging and when a ring is initially inserted (or replaced because many come out!) it causes stress to the pig. It is perfectly natural for a pig to dig and root the ground and while some pigs with shorter snouts (Kune-Kune, Middle White and Potbellied) might dig less than most it would be misleading to say that there are any breeds that never dig.

If you have an immaculately lawned area or pasture you want to preserve then the only way to deter your pig from digging is to put a ring in its nose. However, in this instance it may be better to consider whether you should have a pig at all. The pigs I keep are not ringed and are allowed to do whatever comes naturally to them. They get iron and minerals from the soil and I am sure that they benefit considerably from not having a ring in their nose. If you are extremely lucky having rough land or woodland in which your pigs can run freely then this issue may never arise. It does not matter to us how much rooting they do or how much they churn up the land and it will probably be a great advantage because this may be one of the reasons you got your pigs in the first place. Pigs make excellent rotovators.

Ringing your pig is not a difficult task but I suggest you think very carefully before taking the decision because I am sure that any trust that is built up between you and your pig will be detrimentally affected if you decide to take this route.

Insight: Moving pigs – a battle of wills

A pig is very strong and desperation certainly set in the first few times I tried to move and/or collect older pigs. The mistake I made was touching them at all. I soon learned, thanks to a very experienced pig keeper, never to touch or try to push them to get them to go anywhere. Use a bucket and feed and they will follow; trying to push or pull them will send them in the opposite direction.

10 THINGS TO REMEMBER

1 *Never get a pig ark that is too big for your needs. Not only is it a waste of money but also pigs are very clean animals and will never dirty their beds. If, however, they have too much room they will turn it into an en-suite!*

2 *Don't leave getting everything your pig needs to the last minute. When your pigs arrive you want to spend time enjoying the experience not driving madly around trying to get what you should already have in place.*

3 *Barbed wire at the 'bottom' of your fencing and reversing the hooks on the gate are two important fencing issues that you need to make sure are done.*

4 *There is no such thing as a 'grass eating, keep your lawn immaculate' pig. Yes – all pigs eat grass but all pigs also dig and root, and although some do this far less than others digging and rooting is a natural pig activity.*

5 *Make sure that the pig housing (ark) you have faces away from the prevailing winds. Pigs do not like draughts.*

6 *Don't put feeding troughs and/or water bowls on the fence line or by the entrance gate. It might make feeding/watering easier but the pigs will gradually undermine the fencing digging for spilt food or will block the entrance with rotovated soil.*

7 *Train your pig with electric fencing if they have not been used to it. It only takes an hour or so to educate them.*

8 *If you are not experienced in erecting stock fencing it is often cheaper in the long run to pay your local farmer or professional fencer to do it.*

(Contd)

9   *Water is heavy so you don't want to carry it far. Make sure you have a tap close to the pig pen so you can attach a hose and save unnecessary water carrying.*

10 *Think how nice you feel when you go to bed at night in nice clean sheets. Pigs are no different so put clean straw in their arks at regular intervals.*

4

Can keeping pigs be cost effective?

In this chapter you will learn:
- *about the financial side of keeping pigs*
- *about feeding pigs for different purposes*
- *how much you should expect to pay*
- *the most common mistake of them all.*

This is the sort of question that everyone asks but is one that only you can answer! You must of course start by asking yourself whether you want your pigs to be cost effective or not.

Many people believe that the pleasure they get from keeping their pigs is return enough if they are keeping a pig as a pet. Many who keep pigs to fatten for the freezer believe the benefits gained from having real meat (tasting just like pork should taste) available all the time more than compensates for the cost of keeping them. Even many people who keep rare breed pigs to breed get more pleasure and satisfaction from knowing they have done something to help save the breed than they do from any monetary reward. While these views are very laudable and will, in many cases, be more than enough to justify becoming a pig keeper it is important to realize and understand the true cost of keeping a pig. It is not fair to your pigs if you find that, because of your failure to make adequate financial plans, you cannot provide and care for them in a proper manner. While, in my mind, it goes without question to say that 'happiness is pig shaped' you must be aware that this 'shape' can soon rapidly change if your pigs start eating you out of house and home. It costs money to become a pig keeper and even more money to remain one.

Your initial outlay

You will be happy to know that you do not need to be a millionaire to become a pig keeper. Provided you have the land and it is adequately fenced then your initial outlay is limited to the cost of your pigs, bedding, water bowls, feeding troughs and any expense you might incur in organizing and collecting the new additions to your family.

Cost of pigs

This depends on so many factors that it is almost an impossible question to answer but if you are aware of the factors that can affect the price of pigs then you will be able to make a reasoned judgement. The ten factors that can affect the price you might pay for your first pigs are as follows:

1 Age *A piglet or weaner (approximately eight weeks old) will obviously cost you less that a fully-grown pig. As a new pig*

keeper you should be starting with the youngest pigs possible so that you can build up a relationship with your pigs during those early weeks and months when you are gaining experience. Be sensible and avoid jumping in at the deep end by getting a fully-grown pig as your first experience of the porcine race.

2 **Sex** *A female (gilt or sow) is usually more expensive than a male (boar or hog). 'Gilts for breeding or for producing bacon' and 'boars for pork' is not bad advice to follow. Castrated boars make the best pets if you are not going to breed. Entire boars should only be part of the equation if you are going into large-scale breeding and you will need to have two or three years of good pig-keeping experience before you take this major step.*

3 **Registered** *A pure bred registered pig accompanied by all the correct paperwork is more costly than an unregistered pig. If you are not going to breed or show your pigs and are keeping them just as pets or for fattening you have no need to incur the extra expense of registration.*

4 **Breeding, fattening or pets** *You will generally pay more if you are buying for breeding than if you are just keeping a pet or rearing for meat. I believe that breeding should be the second step to take and not the first.*

5 **Breed choice** *Prices can vary depending on your breed choice although unless they are extremely limited in numbers (Mangalitzas) or have a high 'pet' profile (Kune-Kune) the difference in price between one breed and another is not great. Commercial 'pink pigs' that have been bred and reared under an indoor commercial environment are not as hardy as the old, traditional, outdoor breeds.*

6 **Markings, underline and general conformation** *These factors are generally taken into account if you are looking for a 'show pig' and can affect the price. As a new pig keeper you are unlikely to be treading this path at such an early stage because you will have enough to learn about basic pig keeping without concerning yourself about ring etiquette, breed standards, conformation, underline and breed lines. Not to mention the time you will have to spend training your pigs in ring craft in preparation for the big day.*

7 **Castrated or not** *If you have a male pig that you are not using for breeding and ask the breeder to have it castrated for you then this extra expense incurred will usually be added to the cost of the pig.*

8 **Rare breed or commercial** *You should expect to pay a little more for a traditional rare breed pig or piglet compared with a commercial standard pink pig or piglet of the same age. As with most things in life, you get what you pay for.*

9 **Livestock market, pig sale or private** *While you might pay less for your pigs if you buy through a livestock market you would be better, during your early days while you are learning what to look for, to buy privately from a good recognized breeder. In this way you will be able to see the mother, the conditions that the pigs are kept in, how they have been reared and also seek the advice and help of the breeder in making your selection. Knowing the provenance of the pigs you are getting and having access to the breeder if you have problems is very beneficial to any new pig keeper. You will also have direct access to all the required registration certificates and breed history if you are buying pure bred registered pigs.*

10 **The breeder** *You should not be surprised that the better the breeder and the better the pigs are cared for then the more expensive they are likely to be. It should also go without saying that in the longer term it is better to pay a little more for pigs with accreditation that you know are healthy and have been reared well in good conditions. The alternative of paying less for any pigs that have been kept in less than ideal conditions and raised as cheaply as possible is not a road I advise any new pig keeper to go down. You will almost certainly end up regretting your decision as you watch your pigs grow less despite eating more (were they wormed?) or watch any savings you might have made, by buying on price alone, disappear completely as you pay the inevitable veterinary bills that will come your way if your new purchases need attention. There are no controls or restrictions on pig prices so breeders can charge exactly what they feel is right or in some cases whatever they feel they can get.*

These ten factors should act as a guide when it comes to buying your first pig but in general terms buying to fatten is cheaper than buying to breed while buying a 'pet' pig is somewhere in-between. When considering the price a breeder may be asking for a weaner piglet consider the following:

▶ *A sow or gilt that is used for breeding will eat around 1,000 kg of pig pellets a year.*
▶ *The breeder will have devoted around 365 hours a year to looking after, cleaning out and feeding the pig even if no longer than 30 minutes both morning and evening is spent in animal husbandry. Not a long time each day with a breeding sow but it mounts up over a year.*
▶ *The breeder will have had to pay all the veterinary expenses for vaccinations, worming and general animal health care during the year.*
▶ *The breeder will have certainly have had the expense of either 'boar hire', artificial insemination or will be feeding an 'in-house' boar in excess of 1,000 kg feed a year.*
▶ *The breeder will have paid for advertising the weaner piglets for sale.*
▶ *Birth notification and registration with the BPA may also have been paid for.*

All the above, without taking into account costs such as bedding, light, heat and general maintenance (fencing, housing, etc.) are also costs that you will incur if you become a breeder yourself. Do not even start to think that the price you might be asked for your first weaner piglets is all profit to the breeder because as you can see it certainly is not. If you are ever tempted to think that pure bred, rare and traditional breed piglets are expensive just compare the price the breeder asks with the current price of a pedigree puppy, kitten, alpaca, llama or ostrich. You will be very pleasantly surprised to find that starting your new pig keeping venture is far less expensive.

By far the best thing to do is to visit the breeder as early as possible after the piglets are born (so you are first in the queue) and ask the prices of the individual piglets you like. A good reputable breeder will always help and guide you and unless the pigs are very rare or there is some other special reason for a high price tag you will rarely find even the most excellent of rare breed registered pedigree weaners expensive. The pig was always a favourite of the smallholder or 'cottager' because of its reasonable cost and the same situation remains today.

Insight: Just get on with it

Good advice comes in various forms and when I was thinking of getting my first pigs I spent a lot of time going from pig keeper to pig keeper asking how I could learn more about pigs. The best advice I got was short, sharp and to the point – 'Get some!'

OTHER COSTS TO THINK ABOUT

▶ **Bedding** *is not hugely expensive at all. Furthermore, if you just have two or three pigs then five or six bales of straw will last you a long time during the summer months if you keep your pigs outside.*

▶ *You will be able to get* **water and feed troughs** *from your local agricultural merchant if you want to buy brand new equipment but finding an alternative to this is often more fun. Local farm sales are well worth the occasional visit and will often save you money.*

Fattening pigs for the freezer

If you are keeping pigs to produce meat for your family, friends and relatives then you need to put on your business hat to make sure that you do not spend more on keeping or rearing your pigs than the return you get. This includes taking into account the time you spend feeding and looking after your pigs.

Some people will tell you that there is no way, under the current agricultural climate, that pigs can be profitable, while others clearly more than cover their costs. A lot depends on how you keep your pigs, how (and at what price of course!) you sell them and how much value you place on your own time in looking after them. I have always believed that pigs (especially the traditional breeds) are undervalued as weaners and certainly undervalued as porkers if they are 'free range' and 'additive free'. Too many breeders sell in fear and regret at leisure.

Insight: No room for the car

I once became the owner of SIX full size, chest deep freezers in my garage until I came to my senses and insisted that my customers collected their meat the day it came from the butchers rather than freeze it. It was much better and healthier too, and saved me an absolute fortune in electricity.

If you aim to sell your pigs at the local market then you can expect to make a loss. You must either accept it or question why you are keeping pigs anyway. You are at the mercy of so-called 'market forces' that inevitably turn out to be professional dealers who will probably make more in the day than you will make having looked after your pigs for months. You cannot expect to get a 'premium price' for your undoubtedly superior pigs unless you manage your business yourself rather than put it in the hands of a dealer or auctioneer.

Coloured pigs of any breed are often discriminated against by the average dealer, butcher and market middle men, and even the factory-farmed white pigs do little better. There is virtually no market for coloured weaners and coloured 'porkers' at local markets. The low prices paid for rare and traditional breed pigs at an open sale or market often stem from the fact that:

▶ *abattoirs find them harder work because of their thicker, hairy coats compared to the commercial pigs whose virtually hairless coats are so much easier to steam clean*
▶ *butchers are aware that many rare breed and traditionally reared pigs have a greater fat layer than they get from a commercially reared pig. They therefore feel they have to make a price allowance for trimming off excess fat or they will end up throwing their profit in the waste bin*

▶ dealers at markets have to take into account the above two factors and the difficulty they might have in selling their purchases to butchers and abattoirs.

What people often fail to take into account is the enhanced flavour, quality and texture that is part and parcel of a naturally fed and reared rare or traditional pork pig. If you can produce pork with 'crackling that crackles like it used to', joints of meat that do not shrink dramatically when they are put in the oven and bacon that does not 'swim out of the pan' as you cook it (not an unusual sight as a result of the excess water in much of the bacon available) then you will have relatives, friends and neighbours beating the way to your door in order to be first in the queue for the 'mouth-watering eating experience' that you are producing.

Insight: The cupboard was bare!

Producing pork and bacon that 'tastes like it used to' is rewarding both to the taste buds as well as to the bank balance but I very nearly missed out on the 'taste buds' side of things. When I produced my first pork the demand for it was so great that when I went to get some for myself from the deep freezer there was none left. From that point on I have always put aside bacon and pork for my family BEFORE I let other people have any.

Even specialist 'rare and minority breed' sales do not always achieve much higher prices, although some excellent promotional work by the British Pig Association is improving matters. If there is nowhere you can confidently take your pigs that will result in you getting a reasonable return for your efforts then this points the way to you becoming a salesperson as well as pig keeper. If you follow a few basic rules and discount your own time you will make your pigs more than pay for their keep.

As a small-scale pig keeper you will probably start by keeping two or three pigs for the sheer delight of knowing what you are eating, what the pigs have been fed and how they have been kept. You will certainly want at least half a butchered pig in your own freezer in order to benefit from all your efforts and enjoy that 'mouth-watering eating experience' yourself. This leaves just five half pigs that you need to find customers for. With family, friends, neighbours and work colleagues undoubtedly wanting to share your eating experience finding just five customers will not be difficult. Managing the marketing and sales of three pork pigs is hardly like being the chief executive officer of a multinational company but you still need to make sure that you are just as professional in your approach.

Insight: Misleading statements

After many years of keeping rare breed pigs I sometimes forget that not everyone talks the same 'piggy' language as I do. A recent incident brought it home to me that I should always remember that what is normal to me is not the same for my customers. I only sell 'half pigs' not joints or slices of bacon, etc. but when a new customer came accompanied by three burly friends 'to help him carry the pig' I realized that he thought he was buying it whole and would have to cut and joint it himself. I shall make things clearer in future!

If anyone tells you that your piglets, breeding stock or pork is overpriced they are wrong. They may be able to buy cheaper elsewhere but this produce is not your stock, your meat or your quality. Anyone buying your pigs is getting a good deal based on the true costs of production, especially if they are free range, rare and minority breeds. They are also getting a quality animal that will produce meat that tastes like pork should do and not like cardboard.

THE COST OF FEEDING YOUR PIGS FOR FATTENING

Most new pig keepers do not realize that the optimum age to send a pig to the abattoir if you are producing pork is 26 to 32 weeks provided it is living outside and is being fed naturally. This means that the first thing to take into account is that if you are keeping your pigs to produce pork for the freezer you will only be keeping them for between 18 and 24 weeks because they will be around eight to ten weeks old when they arrive with you. If you are producing a pig for curing so that you have bacon, gammon and ham then you will keep your pigs for a further two to four months or until they are about ten months old.

Insight: It would have been a miracle

It took an old (both in age and experience) and well-respected pig keeper in the 1960s to put me right many times in my early days. I remember well how he laughed when I said I was going to have half the pig as pork and the other half as bacon. 'How's that?' he said. 'Did you not know that the best pork comes at 26 weeks of age and the best bacon comes from a sow that's had one litter?' He smiled a knowing smile and simply asked, 'How is the second half going to run around so you can have your bacon when you've had the first half for pork?' Oh! How I loved that man!

Working out the approximate costs of feeding your pigs is easy if you use the following 'rules of thumb':

▶ *A pig should be fed about 1 lb (0.45 kg) a day for each month of its life from weaning up to six months of age.*
▶ *At six months of age you should stop increasing the feed.*

This is slightly more than you will actually feed your pig but it is a very good guide that is easy to remember and is perfectly adequate for you to work out the cost of feeding.

Simply find out the cost of a 56 lb (25 kg) bag of food and calculate accordingly. At two months of age your pig will eat about one bag a month, at four months two bags and at six months three bags. If you are keeping a 'pet' pig such as a Kune-Kune or Potbellied pig then this 'rule of thumb' does not apply as they require less food because they are smaller and may have a slightly different diet.

If you feed waste fruit and vegetables to your pigs, within the DEFRA rules and guidelines, then you will need to reduce how much concentrated feed you give to your pigs accordingly. But remember that your pigs need to have a balanced diet to give them the required level of protein intake to enable them to grow steadily and normally.

Feeding your 'pet' pig

If you are planning to keep one of the much smaller 'pet' pig breeds such as the Kune-Kune or Potbellied pig then you will not need to feed either as much food or such high levels of protein. During the summer months Kune-Kunes thrive on grass and a small amount of pig feed that has a low protein level, but during the winter, when there is a lack of good fresh grazing for them, they will need more of the concentrated pig feed. There is a special Potbellied pig food available on the market that is ideal for all 'pet' pigs as it is correctly balanced for their needs and has a lower than normal protein level (12 per cent).

Insight: Say little and learn a lot

I have been grateful to many people who have helped me along my pig-keeping path and when I was first starting up I soon learned that listening is better than talking and pretending I knew more than I did. At my first Rare Breed Committee meeting I asked an older, much more experienced member of the committee how he thought I should play it. 'Tony, my friend,' he said, 'you can sit there all afternoon saying nothing and let the others think you are a fool or you can say something and prove you are!' Needless to say I said very little.

Feeding a breeding sow

If you feed your sow 6 lb (2.7 kg) a day for 365 days a year she will eat around 40 bags of sow breeder concentrated pellets. This is probably more than you will feed her because although she will need more after she has farrowed, she will have less, perhaps around 5 lbs a day, for much of the year. If she has good pasture, fruit and vegetables and other extras it will probably be even less.

The following formula will give you an accurate picture of your operating costs and will enable you to work out the correct price at which to sell your piglets.

Expenditure (insert current prices)

40 bags of pig feed _____

Hire of boar or artificial insemination (twice a year) _____

Bedding, straw, etc. _____

Injections and veterinary services _____

Advertising weaners twice a year _____

    Total expenditure _____ £ _____

Total piglets born and reared in the year _____

Total expenditure ÷ total number of piglets = minimum average selling price per weaner to 'break even' _____

This calculation is very basic and does not include any allowance for your time, capital expenditure you may have incurred on water bowls, feeding troughs, housing and fencing that you may or may not wish to include. When you have calculated your 'break even' selling price you will need to add on an additional percentage for your 'profit' so that you get some return for your time and effort.

Neither selling nor buying pigs at cheap, non-cost-effective prices does anything to help save our rare breeds. If you really care about pigs and the rare breeds then you must be realistic about the costs involved.

Remember that keeping pigs should be fun, will probably be therapeutic, will certainly be emotionally rewarding and will give you hours of pleasure (with a little frustration thrown in!). Sadly all this can, in some cases, go out of the window if you do not at least cover your costs so make sure you do the right thing and charge the right prices for your pigs. Not everyone can afford to subsidize the cost of keeping pigs and make a loss (although some do so very happily). Happiness is sadly not always 'pig shaped' as it is so often shaped like a monetary coin. Make your new pig keeping venture both happy and cost effective.

Probably the most costly mistake of all frequently comes about as the result of success. As strange as this might seem it is a fact that the early 'success' so often experienced by new first-time pig keepers who follow all the simple basic rules time and again leads to their longer-term failure.

There is a very big difference between having a few weaners to fatten for pork/bacon and having four or five breeding sows to sell piglets and/or rear all these piglets for meat. Regularly I come across very nice people who should know better trying to run before they have even learned to walk. Success with a few weaners or a single breeding sow does not mean that you automatically increase your success (and with it your profit) in direct relation to the number of pigs you have. What is achievable with a few pigs is not always achievable with a lot of pigs and if it is then it is definitely not achievable overnight.

The calculator is a wonderful tool but used foolishly it can be your greatest enemy. Do not think that if you made a small profit from breeding and selling a few weaners from one sow that all you need to do to make ten times that profit is to increase your breeding sows by tenfold. Do not think that if the three weaners you bought and fattened for the meat market gave you a good return, then you would make ten times the profit if next time you bought 30 to fatten! In fact, many people have already moved on to calculating that they could be a millionaire by a week on Wednesday if they had 3,000! It simply doesn't work that way.

Insight: All profit – no pig

The first profit I made from a pig came from a pig I never really owned. It was at a Rare Breed Survival Trust Sale. I had just successfully bid for a Berkshire sow when I got a tap on the shoulder, 'I'll give you £25 profit on that' a voice said. It turned out that he had been held up and missed the sale of the pig he wanted. Twenty-five crisp pound notes later (no pound coins in those days!) he became the owner of a Berkshire sow that I had only 'owned' for about three minutes!

Pig keeping can be great fun and should be very rewarding both emotionally and financially but you must build slowly and at each stage of growth make sure that you have your correct sales and marketing plans in place. One step at a time is the rule and then only take that step if you are really on secure ground as far as your future sales are concerned, whether it is selling piglets or selling meat you are interested in. Finding five customers for your pork and bacon is far easier than finding 50 and securing orders for ten piglets is much easier than finding buyers for 100!

Insight: Happier and healthier

The benefits and gains you get from keeping pigs are not only financial. I took early retirement due to ill health that was brought about by stress caused by the job I did. My pigs offered me a much more sensible lifestyle because they are not interested in budgets, forecasts, expenses, targets, profit schedules, next year's forward planning and the competition.

10 THINGS TO REMEMBER

1 *Buying cheap does not mean buying well. Quite often buying cheap means buying trouble.*

2 *Always buy from a recognized breeder who you can go back to if you have problems.*

3 *Buying from markets where you do not know the background or provenance of the pig/s you are buying is a bad mistake.*

4 *Initial outlay is less important than profit. A good, well bred pig from a recognized breeder will always cost more but will give you a better return both financially and emotionally.*

5 *Go into keeping pigs with your eyes open and your wallet shut. You should only open your wallet when you know that what you take out is going to come back with interest.*

6 *Keep a note of your costs. Pig keeping can only be 'cost effective' if you write down everything you spend. Selling anything (pork, bacon, weaners or adult pigs) for less than they cost to produce will make you a loss and send you bust.*

7 *Learn how to make your pigs (not you) pay for everything.*

8 *It is only by making your small-scale pig-keeping project profitable that you can give your pigs the sort of life they deserve.*

9 *Never buy out of sympathy either for the pigs (currently badly cared for or in awful conditions) or for the owner (desperate to get rid of them with a thousand sob stories). You will only buy trouble.*

(Contd)

10   *Buying cheap is as bad as selling cheap. It is not clever to buy a piglet for less than you know it costs to produce because you are aiding the demise of the breeder (if they go out of business where do you go next year?) and YOU are helping set a market price far less than it should be (if you buy at that price how can you justify selling at a higher price when you have effectively said that you paid the right price?).*

5

Getting your first pig

In this chapter you will learn:
- *how to select the right breed for you*
- *what to look for when buying a pig*
- *about the different breeds you should consider.*

Insight: I love them all

I must admit that when it comes to selecting my favourite breed I am a bit of 'Numpty Head' because I love them all.

What breed should you start with?

If you were to ask ten different breeders this question you would probably get ten different answers. All breeders have their personal favourites but it does not necessarily mean that their favourite is the right breed for you. In time, when you have some experience of the different breeds, you will find a favourite of your own but this will be a very personal choice based on your own experiences.

Insight: The very special one

I once fell so much in love with a little British Saddleback piglet I was rearing for pork (26 weeks) that I decided to keep her longer for bacon (11 months) but when that time came I changed again and decided to let her have one litter

(Contd)

(18 months). Nobody who knew how much I adored this pig seemed at all surprised when we recently celebrated her 15th birthday, now happily retired and living in the woodland. Happy Birthday 'Flossie'!

At this stage in your pig keeping life there are four main factors that should determine the breed that you start with:

1 *Whether you want a 'pet' pig, a pig for meat or a pig to breed from.*
2 *The ability to locate a good breeder.*
3 *Availability in your area at the time of purchase.*
4 *The pig that visually appeals to you.*

If you are lucky then you may be able to find a pig that meets all four of the above criteria but all too often you will find that you have to compromise on what would have been a perfect starting point. If you cannot find a good local breeder that has stock available when you want it then you may have to drive further to get your pigs than you had originally planned. It is better to drive a little further to find the right pigs and the right breeder than it is to accept second best.

Although you may find what you are looking for in the small ads section of your local paper you are more likely to have success by looking for breeders in the more specialist farming and smallholding publications. If you have access to the internet then you will get a wealth of information about breeders and stock that is available by visiting the British Pig Association website (www.britishpigs.org.uk). You will be able to search by county, by region or indeed the whole country for breeders who currently have pigs for sale, all from the comfort of your own home.

Insight: Children and pigs

In my early days, because I had children to consider, I looked at all the breeds hoping to find the smallest possible to breed for pork. How stupid I was. Because they are only 26 weeks of age when they go for pork they are ALL small no matter what breed.

What to look for when buying your first pigs

Although you may think that line and form (i.e. the pigs conformation) play the most important part in selecting your first pigs, this is not necessarily the case. These factors are of paramount importance to the more experienced pig keeper or to someone entering the show ring but there are far more important things for you to consider as a new pig keeper. If you look for the following things then you will have the basis for ensuring you obtain good healthy weaners to start you on your way.

Insight: Beware of the dealer

I once met a pig keeper who had piglets of six different breeds and yet no sows or boars! His pig pens were spotless and I was very impressed until someone told me he was a 'piglet dealer' a bit like a 'puppy farm' that has no dogs. It made me very wary of buying any piglets from anywhere I could not see their mother because buying a pig without some provenance is not wise.

1 *Look around the premises as you park your car. Is it reasonably tidy and has an attempt been made to have some order about the place? If there are bits of glass about the place, barbed wire, old cans and buckets, scrap metal and discarded junk about the place beware because this is hardly someone who is likely to care about stock. No farm or smallholding can be spotlessly clean or pristine because with animals, tractors and farm vehicles about there is bound to be mud and a certain amount of dirt. The good stock person will realize the dangers of an untidy, ill-kept yard and do something about it. If you are happy, move on to meet the pig keeper.*

2 *As you enter the farm or smallholding where the piglets have been reared, slow down and take a careful look at any other stock on the premises. Look at the sheep to see if they are limping or in poor condition, the cattle to see if they are a bag of bones or have that 'please feed me' look and at any horses to see if they have that happy and contented look about them. How all the other animals are kept on the farm will be a good*

indicator as far as the general animal husbandry standards on the premises are concerned. If you are not happy with the other stock then remember that the piglets you are considering buying have been brought up under the same regime. They may look good in the nice, clean, newly strawed show pen they have been put in for you to see but beware because the real story is reflected by the condition of the other stock.

3   How the pig keeper greets you and how they treat you is very important indeed. If the pig keeper rushes the meeting or greets you by saying something like, 'They are all in the end shed and are all for sale so go and take a look. Just pick anything you want and let me know' it is clear that he/she is more concerned about selling pigs than really finding out what is best for you. If the pig keeper is really going to be of benefit to you and help you get what is right for you then you will be asked questions such as:

▷ Have you kept pigs before?
▷ Are they to be kept inside or outside?
▷ Are you going to breed, are you fattening or simply keeping a pet?
▷ Do you want the pigs registered or not?
▷ Have you got young children? (The pig keeper can advise on the most suitable pig in terms of size, sex and temperament.)
▷ How much space have you got to keep the pigs in?
▷ When do you want the pigs?
▷ Have you got your housing, fencing, straw and feed all in place?

Questions like these will demonstrate that the pig keeper is interested in what you want and not just in selling you anything that might suit him/her.

I believe it takes about an hour to select piglets successfully and if you are happy with the way that you are greeted and treated then now is the time to move on to looking at the pigs.

4   Although the pig keeper might offer to take you to see the piglets you should make it clear that if the piglets have been weaned, you would like to see the mother (and if possible the father) before you see the piglets. Seeing the mother, with or without the piglets, is all important because piglets always

reflect the qualities of the parents. If you are looking for a small 'pet' pig and the mother is a giant, then the piglets you buy will be of a similar size. If the piglets have been separated and put in a nice clean 'show' pen and the mother is in a pen that is thick with muck that also tells you something. If the mother is separate from the weaners check her ear tags, notching on the ears or tattoo number and mentally note the last three numbers. You will need to compare this later with any paperwork to make sure that you have actually seen the right mother. The piglets might have been brought in from a sale and the sow you saw might have nothing to do with them.

5   *If you are happy with the sow and the piglets have been weaned you can move on to see the piglets. If the pig keeper has asked you all the suggested questions he or she will probably enter the pen and immediately mark three or four piglets that meet your requirements. If you are breeding they will mark the gilts, if you are fattening they will mark the boars, if you want registered weaners they will mark all those that meet the breed standards and can be registered. This will save you a lot of work because out of 12 piglets perhaps only three or four will meet the requirements you are looking for.*

6   *Having marked those piglets that are suitable for you then be prepared for a surprise – the pig keeper may ask if you have any questions before going back to the house, leaving you alone with the piglet. This is a good thing because the pig keeper realizes that he or she has given you all the advice that you require and now the choice must be yours. As you have done the right thing and gone to a recognized breeder while also having taken the sensible steps listed above you are able to make the final decision yourself.*

7   *Return to the house where hopefully the person selling the pigs has made you a cup of tea and then you can complete the paperwork. If you are buying registered pigs make sure the ear number you saw on the sow matches the number on the registration form. If you are taking the piglets with you, make sure you get a Movement licence given to you by the seller. If you are collecting at a later date, pay a deposit to secure the order (you don't want anyone else buying your pigs) and enjoy the moment of knowing you are a pig owner.*

Rare and traditional breeds

The main traditional, rare and 'pet' breeds that you have to select from, all offer something special to the new small-scale pig keeper that may not be found in the more commercial breeds. If you obtain a weaner piglet from a commercial pig unit where all the sows and piglets are kept indoors your piglet may not have the same hardiness necessary to cope with living permanently outside as a 'rare breed' or traditional pig breed. Although one of the more commercially farmed 'pink' pigs may be cheaper to buy than a 'rare breed' or traditional breed of the same age you may find that it does not have the appeal of its more colourful relatives.

BERKSHIRE – A FANTASTIC PORK PIG

The Berkshire (see Plate 4) is a truly great pig with impressive pricked-up ears. A black pig with white blaze on the face, four white feet and a white tip to the tail similar to the markings you would expect to find on an Oxford Sandy & Black. A pig of great personality and very good temperament who has the sort of mothering qualities that makes farrowing and bringing up piglets a sheer delight.

The Berkshire is a traditional pork pig that produces some mouth-watering joints and chops with 'crackling' that is second to none. The excellent carcass quality made it an early favourite with the Royal Family who for years kept a large Berkshire herd at Windsor Castle. The first Berkshire pig ever recorded was a boar called 'Ace of Spades' and was bred by Queen Victoria.

The original Berkshire was, in appearance, much closer to the Oxford Sandy & Black, being reddish or sandy coloured and sometimes spotted with black. This stock was later refined with a cross of Siamese and Chinese blood which then set the colour pattern we know today. It is said that the Berkshire was discovered by Oliver Cromwell's army over 300 years ago in their winter quarters at Reading.

BRITISH LOP – AN EXCELLENT MOTHER

The British Lop (see Plate 8) is one of the most endangered of all our native breeds. It is extremely hardy and can be kept outdoors all the year round. They are excellent mothers, very docile and the breed is well recognized as having a very good record as far as piglets born and reared is concerned. A good long pig that is renowned for producing excellent pork and first-class bacon. Originally found in the Southwest of England in Cornwall and particularly around the Tavistock area of Devon. The British Lop is closely linked with the Welsh and also the now extinct breeds of Cumberland and Ulster. An all white pig with large lop ears which until the 1960s was known as the 'National Long White Lop Eared' pig. Being an all white pig it suffers none of the sales and marketing disadvantages experienced by the coloured and spotted breeds when being reared for meat and sold through commercial outlets.

Insight: Are some pigs more docile than others?
For many years I believed the story that 'lop eared' pigs were more docile than 'prick eared' pigs. Not true. They just seem more docile because they can't see you coming!

BRITISH SADDLEBACK – A GREAT DUAL PURPOSE PIG

The British Saddleback (see Plate 6) is the result of combining the herd books of the Wessex Saddleback from the New Forest area of England and the Essex (excluding the word Saddleback) of East Anglia. The two breeds combined in the 1960s because numbers were dropping so fast. The two breeds were different in appearance with the Essex having the broader white saddle and four white feet while the Wessex had broader hams and ears that tended to tip more forward. In other ways they were very similar and the combination of the two breeds has resulted in an excellent pig, ideally suited to being reared in a natural outdoor environment. With semi-lop ears, a long snout and a distinctive white 'saddle' across its shoulders, this predominately black pig is unmistakeable in appearance. This is great dual purpose pig, renowned for both its pork and bacon.

GLOUCESTERSHIRE OLD SPOT – THE 'ORCHARD' PIG

Probably one of the best known of all the British rare breed pigs, the Gloucestershire Old Spot originally came from the Severn River valleys in Gloucestershire. The breed lived mainly on a diet of the whey from the Gloucestershire cheeses and apples from the very prolific orchards in the area.

A lop eared white pig with black blotches and spots, it is a good dual purpose (pork and bacon) pig that is both hardy and docile. This is a pig that is very much a smallholders' pig and was known in its early days as one of our 'orchard' pigs. With good mothering qualities the Gloucestershire Old Spot's attraction to first-time pig keepers can be well understood.

A wonderful pig that many say would have been far fewer in numbers had it not been for one man, George Styles, who in some circles is affectionately known as the 'grandfather of the breed'. Styles was a prolific breeder of the Gloucestershire Old Spot and developed the 'cyclical breeding system', which is still being used by many breeders today. It is in no small part thanks to his passion for the breed that the breed survived and flourished at a time, when without such support and dedication, we might have found ourselves with far fewer blood lines than we now have. Very sadly George Styles passed away in 2009 leaving behind very many 'Gloucestershire Old Spots' pig keepers who owed him a great deal for everything he did for the breed.

Because of the difficulties of selling spotted and coloured pigs to butchers (it was claimed that cleaning them was harder and that the customer did not want such spotted pigs) there was a tendency to breed pigs with as few spots as possible. Thankfully now this is a trend that is being reversed. The popularity of this breed remains as high as ever and it is often easier to want one than actually find one so make sure that you find a suitable breeder and place your order early.

LARGE BLACK – THE 'ELEPHANT EARED' PIG

If you like lop eared pigs then you won't find much better than this all black pig that some pig keepers affectionately call the 'elephant eared pig'. With its fantastic lop ears the Large Black (see Plate 5) has probably acquired this nickname because of the similarity the newly born piglets have to a very small black elephant. Viewed from behind, after just being born, their huge ears and little straight tail certainly make one think of a baby elephant.

Many stories surround the arrival of these wonderfully long, lop eared pigs on our shores. Historically, an all black lop eared pig was not known in this country until some time after the arrival of Siamese and Neopolitan breeds.

One possible way in which these pigs arrived in this country and a story that does seem to have some credibility revolves around the need to keep sailors free of scurvy. It is said that when Chinese trading boats were making the long and tortuous journey to our shores they would always ensure that they had plenty of livestock on board to feed the crew in order to help prevent scurvy. In the days when ships relied on the prevailing winds they could never be certain how long the journey would take or how long they would be at sea.

On this particular occasion it is said that two Chinese trading vessels made very good time with both landing early. One docked in Plymouth and the other at the Thames basin outside London. A large number of surplus black pigs were supposedly unloaded at both places. The farmers in both East Anglia and Cornwall took a liking to them with the result that in the fullness of time the Large Black of today emerged. When you take into account the fact that the black pigs, specifically the Cornish Black, were mainly found in the South and Southeast of the country as well as in East Anglia it does seem that this is probably more than just a folk tale.

It is certainly true that the big black pigs bred in the Southwest became very popular abroad and is still known as the Cornish Black in many parts of the world. It is also true that black pigs and black pigs with a white saddle were found all around London, within reach of the Thames basin. There is, however, no way of proving the authenticity of the story.

In the early nineteenth century these pigs were described as being one of the largest of all our pigs with very big heads and ears that were so long that they could hardly see which way they were going. This is still very much the case today.

This is, without any doubt, one of the more graceful, elegant members of the pig family. Despite its size, it is very docile and is an excellent mother, capable of rearing large litters as well as producing excellent bacon for the table. Although it is much more of a bacon pig than the smaller, rounder, much chunkier pork pigs such as the Berkshire and Middle White it still produces excellent pork at around six months of age.

> **Insight: Where can I get black crackling?**
> I was 12 months into pig keeping before I realized that my search for pork with black skin was futile because black and ginger skinned pigs (Tamworth and Large Black) kill out white just like any other pig – oops!

MIDDLE WHITE – SO UGLY IT'S BEAUTIFUL

The Middle White (see Plate 3) is a truly magnificent all white pig with a flattened face that is squat and round with its short round body giving it a really 'chunky' appearance. The flat snout gives it the appearance of having just walked into a wall and lost, but the wonderful big pricked ears, with an amazing feathering of hairs around the edges, simply add to the overall appearance of this awesome pig. The Middle White is said to be so ugly that it is beautiful and coming towards you it certainly has the look of a vampire bat.

The Middle White is a really good natured pig that is very talkative and has character all of its own. Originally bred in Yorkshire as the London Pork Pig in order to compete with the Berkshire for the London pork trade it still retains the qualities that produced the great pork of the past. Originally the greater majority of the white pigs came from the North of England while the coloured pigs were usually found in the South.

The Middle White is thought to have come about as the result of selected breeding with offspring from the Large White and Small White (now extinct). All three originated in Yorkshire and were known as the Small Yorkshire, Middle Yorkshire and Large Yorkshire, which is the name they are still known by in parts of Canada and America.

OXFORD SANDY & BLACK – A RECONSTITUTED BREED

Not recognized as a 'rare breed' by the Rare Breeds Survival Trust, this wonderful pig is known as a 'reconstituted' breed. The breed society was formed in the mid-1980s and the foundation stock used could not be verified (other than by word of mouth) as being pure Oxford Sandy & Black because there had been no herd book, birth notifications or registrations for many years. Despite this, the Oxford Sandy & Black (see Plate 9) is a breed that is growing in stature year by year as it is extremely popular among small-scale pig keepers and can make an excellent 'starter' pig.

Because of its history and background the Oxford Sandy & Black does not breed true to type. In any litter the piglets can range from all sandy (looking like a Tamworth) to white or pale cream with black spots (looking like a Gloucester Old Spot) and occasionally almost black (very similar to a Berkshire).

This breed is renowned for the quality of pork and bacon that it produces and, as colour does not matter when it comes to rearing for meat, there is always an outlet for unregistered pigs as 'fattening' stock. Definitely one of the best pigs for a first-time pig keeper who is considering breeding as it has a great personality, is very docile, makes a wonderful mother and is easy to handle. This pig does not seem to have a bad bone in its body and is great where children are concerned.

With black blotches (not spots) on a sandy background, lop or semi-lop ears, white blaze, four white feet and a white tip to its tail, the Oxford Sandy & Black certainly does what the breed society says – it 'catches the eye'. A very hardy pig, the Oxford Sandy & Black will happily live outside all year round with the sow looking after her piglets without difficulty under any conditions.

It is very important to be aware that if you are breeding Oxford Sandy & Black pigs your litters will be a mixture of colours, shapes and sizes (see Plate 9), with very few meeting the strict breed standards required for registration. While having such a variety of colours makes the litter very attractive the only pigs that can be registered are those that meet the strict breed society breed standards. However, even those that cannot be registered make excellent long-term companions or meat pigs. The breed society publish excellent guidelines to help and assist breeders in selecting only those piglets that meet the strict breed standards for registration with the BPA. Pigs not meeting those standards must not be registered for it is only by adhering to the breed standards that the breed will gradually become more uniform in appearance.

TAMWORTH – THE ARISTOCRAT

Regarded by many as the aristocrat of the pig world, this very proud long-legged ginger pig with wonderful large pricked up ears will always command attention.

Although in its present form the Tamworth (see Plate 7) as we know it today is a very distinctive pig with long legs, pricked ears and a pure red or ginger coat, this has not always been the case.

Around 1800 it was said to be much smaller with shorter legs and ears that were far less prominent. As far as colour was concerned it was described as 'spotted red and brown'. Old pictures of the Tamworth show it to have been anything from white or pale ginger with black spots to a 'red and black' pig. It closely resembled the picture painted by Low of the old Berkshire and it was not a million miles away from the current requirements of the Oxford Sandy & Black.

The Tamworth is one of the great dual purpose pigs producing stunningly good pork and equally tremendous bacon. In the mid-1990s the Tamworth came top in a taste test carried out by Bristol University using both commercial and rare breed pigs in a scientifically controlled experiment. It was later suggested that further investigation should take place to establish just what it was that gave the Tamworth meat such a distinctive taste that put it way above all the other breeds.

Although sometimes referred to as 'boisterous' this should not be interpreted as being 'nasty' or difficult to handle. It is a big pig that is full of life and you could never expect the 'aristocrat' of the pig world to be anything other than full of life. A pig with one of the longest snouts, it is of great use to the organic gardener or anyone who has a piece of rough overgrown land that needs clearing. The Tamworth could certainly earn its keep as a professional rotovator.

In the mid-1800s it is clear from records that the Berkshire and Tamworth were closely linked. In view of the association between the Oxford Sandy & Black and the Berkshire, who is to say that all three are not in some way connected.

Insight: They all give pleasure

Although, like me when I first started, you may think that the different breeds are all different, you will soon find out, as I did, that they are all the same – just wonderful! It took me only about 12 months to find this out and I wonder how long it will take you.

'Pet' pigs

KUNE-KUNE

Although not an English breed the Kune-Kune (pronounced
koo-nee koo-nee) is definitely a rare breed (see Plate 1). In 1978
there were reputed to be fewer than 100 left in New Zealand
but thanks to some very dedicated breeders the situation is now
much improved.

Originally a feral pig that lived with the Maori tribes their
name comes from the Maori language – 'Kune' meaning fat and
round. Therefore 'Kune-Kune' in Maori means very fat and very
round. You should bear this in mind when you are considering
the size your pig is going to be when it is fully grown because
many Kune-Kunes will grow to at least knee high and need to be
correctly fed and exercised if they are not to get overweight and
fat. That delightful little piglet you might collect at eight weeks
old is, like any other pig, going to grow steadily in the ensuing
18 months. Not withstanding its natural propensity to grow out
of its charming piglet stage it will still retain all the delightful
characteristics that make it such a tremendous 'pet' pig.

The origins of the Kune-Kune is anybody's guess as it certainly
didn't originate in New Zealand. Perhaps whalers and sealers took
them and released them for food when they visited, maybe the
Maoris brought them over from Polynesia, possibly the settlers
who came to farm brought them or they might just be a product
of a mixture of breeds.

Coming in a variety of colours from cream to black, sandy to
spotted, this is a delightful little pig that does not grow to the size
of normal pigs. Being mainly grass-eating it is easy to keep and
feed. The Kune-Kune is extremely human friendly and very easy to
train but beware because although they may dig and root less that
most breeds and some may not root at all this is something that
cannot be relied upon.

It makes an almost perfect 'pet' pig as it is very docile and adores the company of human beings. Many, but not all, Kune-Kune pigs have a pair of tassels (called 'piri-piris') under their chin very similar to a goat and although the lack of them does not detract from them as a 'pet' they are an important feature if you are thinking of entering the show ring.

Although only a small pig and mainly kept as a 'pet', for breeding or just keeping the grass down, it is claimed by many people that they produce excellent pork.

POTBELLIED PIG

Once the most popular of all the 'pet' pigs in the UK, the Vietnamese Potbellied pig (see Plate 10) now takes second place to the Kune-Kune in the 'pet' pig stakes but if you can find a good breeder who is serious about the breed and then practise good Potbellied pig husbandry you will not find a better, more loyal and giving companion. The Potbellied pig has suffered from a bad press as a result of becoming a 'must have' item in the 70s and 80s with the result that when they went out of fashion they were discarded by many of their owners in favour of the next animal fashion accessory. Too many of these pigs were either overfed on the one hand or neglected on the other and as a result the RSPCA and rescue centres around the country became inundated with them. A 'pet' pig, just like any other pet, is for life not just a whim or fancy to be discarded or its needs ignored whenever the craze fades. The Potbellied pig, with its fine straight tail that constantly wags with pleasure when cared for correctly, is a round bundle of fun with its short snout, prick ears and smooth black, white or black and white coat. This is definitely not a pig that you should keep for anything other than as a 'pet'.

RECENTLY INTRODUCED – THREE IN ONE

In 1972 one of the most amazing of our native breeds, the Lincolnshire Curly Coat became extinct. It was famous for its hardiness and for the fact that its meat ideally suited the land workers of Lincolnshire. However, there was more to this pig than that. In the county of Cheshire was a small factory that made men's sweaters from its woolly coats – it became the only pig that was sheared. This was not the only use that was made of its curly coat because it was prized by fly fishermen worldwide as the finest wool to use when making their own flies because it had a peculiar quality all of its own. In the 1920s the breed was exported to Hungary where it became so popular in Hungary that it won much acclaim at the Budapest show, winning major prizes culminating in the coveted Gold Medal in 1925. The Lincolnshire Curly Coat was so similar to the native Mangalitza that many were cross-bred, producing a pig that was called the Lincolista. Sadly, with the fall of communism, the breed was virtually lost and in 1979 there were only four controlled breeding farms for Mangalitzas in the whole of Hungary. Because of their very close association with the Lincolnshire Curly Coat after years of trying I succeeded in importing 17 Mangalitzas, covering all three colour variants – Blonde, Red and Swallow Bellied – into England in order to establish a breeding programme in conjunction with the BPA (see Plates 2, 11 and 12).

Insight: Keep me warm if you can

Although they used to shear the 'Lincolnshire Curly Coat' pig and make men's sweaters out of them, I have been unable to find any examples. If anyone has an old sweater that Grandad had made from pig wool, please let me know!

Commercial and principally white breeds

LARGE WHITE

The Large White was one of the three breeds originally adopted by the National Pig Breeders' Association when it was formed in 1884. Originally known as the Large Yorkshire it is valued for its carcass both when bred pure and when crossed with other breeds. It is a white pig with very fine hair and is well liked by abattoirs up and down the country because of the ease with which its coat is removed leaving a clear, blemish-free carcass. The Large White is still referred to in parts of the world as the Large Yorkshire and has the reputation of being one of the world's best-known breeds. This pig is currently kept mainly by commercial breeders for meat production. With a slightly dished face and long, pricked and pointed ears, the Large White has a good long back with its snout much longer and more pronounced than its rare breed relative, the Middle White.

THE LANDRACE

The Landrace is a commercial pink pig that, like the Large White, has fine, soft hair that is easy to remove at the abattoir stage, making it the most popular commercial pig in the United Kingdom. The Landrace is now often crossed with a Duroc. It produces a leaner, less fatty carcass than the rare and traditional pig breeds that are reared outdoors. The Landrace is not a native breed and was first imported from Scandinavia in 1949 as a pure bacon pig to meet the demand in this country for the 'Wiltshire Cure' – a process that required a long, lean carcass if it was to maintain its unique appeal. This is a pig easily recognized by its long straight snout, semi-lop ears and very fine white hair.

THE PIETRAIN

This pig was imported into the United Kingdom from Belgium in 1964 and has the classic chunky, rounded look of a first-class meat pig. Although the extra muscles that the Pietrain is known for give it the look of a body builder it is reputed to produce some excellent meat with a lower than average amount of fat. Predominately white with black spots tapering at the edges, a snout that is well defined and broad plus short ears that are semi-lop, the Pietrain is a great pig to use when crossing with other breeds to produce a really good meat carcass.

THE DUROC

Although the hair on this breed is distinctly sandy in colour the skin is white or pink, giving those butchers and abattoirs that reel away from the more traditional coloured breeds exactly what they want. A very hardy breed that makes it a perfect pig for running naturally outside in all weathers. Although the Red Duroc was imported into the United Kingdon from America in the 60s and 70s, a White Duroc has been produced by judicious crossing with white pigs. The Duroc is one of the most popular commercial breeds in America but up to the early 1930s it was known as the Duroc-Jersey. The small head on the Duroc, compared to its body, plus the dished face accentuates its very large, deep chest that is characteristic of the breed.

THE HAMPSHIRE

Many new pig keepers confuse this very special pig with the British Saddleback because of its broad, white saddle. However, although it is certainly related to the Wessex Saddleback, any similarity ends there! The ears are pricked rather than semi-lop and its large, strong skeleton and bone structure leads the Hampshire more towards the role of producing boars for breeding rather than the dual meat role of the British Saddleback. The 'belted' Saddleback was well known in the New Forest in Hampshire, but unlike many other breeds did not adopt the name of the county and it simply became the Wessex Saddleback. The modern day Hampshire is used primarily to produce top-class boars that can then be crossed with other breeds to produce good, lean meat pigs.

THE WELSH

This pig is a fine, white lop eared pig that is fast becoming endangered, which is somewhat of a surprise because it is an extremely hardy pig that thrives in outdoor conditions but is also proven to adapt very well to the more commercial indoor systems. An excellent mother, the Welsh (once called the Old Glamorgan) was revered by the old-time Welsh pig farmers for its excellent carcass quality.

> ### Insight: The right pig for what you plan to do
> Many people think a pig is a pig ... is a pig ... is a pig and I was once one of those people, but I soon learned that each breed of pig has a different purpose in life and different benefits. It was only after I made a few costly mistakes that I realized I had to buy the right pig breed for whatever I had in mind.

10 THINGS TO REMEMBER

1 *Three is the magic number. One loses money, two don't break even but three make a profit.*

2 *ALWAYS ask to see the mother (sow) of the piglets. When I was a young man and I took home my first 'very beautiful' girlfriend my father gave me excellent advice: 'Go take a look at her mother!' Remember to do the same with pigs, because if the mother is a big pig then that is what the piglets will become in the longer term – they will not remain 'Miniature', 'Micro' or 'Handbag' or 'Teacup' sized no matter what you might be told. Genetics determines how we look and grow so if the mother is big the piglets will grow likewise.*

3 *If you can't see the mother be wary. They may be dealers (just like a 'puppy farm' but with pigs) or maybe the condition she is in better reflects how the piglets were really kept before they were put in a nice clean pen for you to see.*

4 *If you want a rare breed registered pig then you must get the paperwork. If they say it is pure bred but they don't bother with the paperwork then all you are buying is 'a pig' no matter what its background. To be pure bred and registered, both mother (sow) and father (boar) must both be registered themselves.*

5 *ALWAYS buy some feed from the seller so that you move them gradually over to your feed from what they are used to.*

6 *Males (boars) and females (gilts) both make good pork but only females (gilts) should be used for bacon or breeding.*

(Contd)

7 *Spend time with the pigs when you go to buy. Don't be impulsive. It may take up to an hour to see, select and finalize your purchase.*

8 *How the farmer/breeder behaves and treats you is as important as the pigs being sold. You might need help and advice later so if you cannot build a rapport when you are buying, what will it be like later?*

9 *It is better to travel a long way to get the breed you want than buy locally and be unhappy.*

10 *The right age to get your piglet is at nine weeks. They should be weaned at eight weeks and sold at nine.*

6

Feeding your pig

In this chapter you will learn:
- *about feeding and growing your pigs*
- *how to combine natural foods with commercial feed*
- *about feed it would be best to avoid.*

When it comes to feeding your pigs you need to remember seven things:

1 *Never overfeed your pig – a fat pig is not a fit, happy pig. Also a gilt or sow that is too fat may be difficult to get in pig, which will cost you money. Pigs will eat more than they should so be strong and do not weaken by giving them too many treats which will end up doing more harm that good.*
2 *Feed a much higher protein diet to young pigs, piglets and weaners to help your young stock grow and develop correctly. There are special feeds on the market that are correctly balanced in content to suit each stage of your pig's life:*
   ▷ *Starter or creep feed (2–6 weeks).*
   ▷ *Weaner grower (6–16 weeks).*
   ▷ *Grower finisher (16–32 weeks).*
   ▷ *Sow breeder (over 32 weeks of age).*
   The above ages of feeding are only intended to be used as a guide and you should always refer to the individual manufacturer's advice as determined by the nutritionist involved in the preparation of the feed. A sow or gilt with

a litter to feed needs to have much more feed while suckling her young in order to assist her in producing the necessary amount of food for her litter.

3 *If your pig does not clear its food within 20 minutes of putting it down you are either feeding too much or there is something not right with your pig. One of the first signs of illness in a pig is when it doesn't eat its food.*

4 *Keep all your troughs, buckets and ladles clean and don't allow the smell of stale food to build up in them. Good basic hygiene is very important at feed time.*

Insight: A body builder? No not me

I never wanted arms like an all-in wrestler so within weeks of getting my first pigs I installed water pipes to all the pig pens along with a full barrel of water outside every other pen. There is nothing heavier to carry than a full bucket of water.

5 *In muddy conditions (remember rain plus pigs equals mud) when your pigs could constantly be turning over their troughs and the feed is getting in the mud you should make a 'feeding platform'. This is not difficult to do and only requires three or four paving slabs. Simply place the slabs in the ground where you are going to feed and put the feed and feeding troughs on these slabs. It is easy enough to brush the slabs clean each day and your pig will happily eat straight from the slabs if they tip over any trough that might be there. You can also put your feed directly on the slabs. In dry weather and on good land, eating directly from the ground is not harmful to pigs as they get lots of their minerals direct from the soil.*

6 *Change the place where you feed your pigs at regular intervals if they are running free outside. If you feed at the same point by the fence each day for a long period of time then it will become a quagmire and undermine your fencing in the winter. It is not always possible but if you can feed away from fencing then do so for it will prevent your pigs undermining your fencing as they dig for any feed that might fall on the ground.*

Insight: My own special quagmire

Running (or perhaps staggering!) out of a pig pen in my mud-covered socks, leaving my wellington boots still standing erect next to the feed trough brings back many memories. It was also a very embarrassing way I had to learn that in wet weather, feeding troughs must be moved regularly. One of the muddiest places in a pig pen is around the feed trough and getting stuck in the mud unable to get out without leaving my boots behind convinced me I should move the feed trough more often!

7 *A pig is what it eats, as is any animal, and therefore you cannot expect your pork to taste as it really should if your feeding regime is wrong. BSE in cattle did not come about by accident and neither was the foot-and-mouth outbreak in pigs simply a freak of nature. Beware of what you feed your pigs and always keep up to date with both EU and DEFRA regulations as far as what can and cannot be fed to your pigs. If you want to produce pork that doesn't shrink and bacon that doesn't swim out of the pan then you must feed as naturally as possible and give your pigs time to grow normally. You then have a 'premium product' that people will want. At the time of writing your pigs must not be fed any of the following:*

   ▷ *Anything at all that has come from a recognized catering establishment or your own kitchen.*
   ▷ *Meat or meat products.*
   ▷ *Fish and products containing fish.*
   ▷ *Eggs or products containing eggs.*
   ▷ *Animal fats (lard, etc.).*

Pig farmers who 'factory farm' aim to get their pigs to a porker weight of about 120–130 lbs live weight in about three to four months. When running outside and fed on a 'non-additive' natural diet chosen from a mixture of pig concentrates, barley meal, fresh vegetables, access to the soil, grass, sugar beet, fruit, bread, excess milk and of course plenty of fresh water, your pig may grow more slowly but will grow naturally at the pace nature intended. It will also almost certainly be happier. It takes around 26–32 weeks to produce the good free range pork that our customers expect depending on the time of year (longer in winter) and the breed being fattened (Middle White and Berkshire pigs finish earlier than most).

Free waste fruit, vegetables and greens from your greengrocer help when looking at your costs as does stale bread and cakes from the bakery. **Never** feed anything that has a meat base or has been in contact with meat (gravy, pasties, pies, etc.). Feeding meat products to pigs is illegal. Diseases like foot-and-mouth can be spread easily via incorrect feeding. Bearing this in mind, be particularly careful of anything from the bakery that may have been in contact with meat.

When feeding concentrates I believe it is much easier and better to feed outside pigs pig nuts rather than have the hassle of making up meal. Being accidentally knocked over by an enthusiastic and hungry group of pigs at feeding time is much less of a disaster if you are feeding pellets than if you are feeding wet meal. If you have to carry your buckets of feed any distance you will soon realize the extra weight you are carrying when feed is mixed with water. However, 'wet feed' does have a distinct advantage over pig pellets if there is not a constant supply of water available for your pigs or if you have to carry water for them anyway.

Insight: Better to feed dry feed

When I sat there in the middle of a pig pen covered in pig swill I learned two things. Firstly never have both hands full with buckets of pig swill when you are surrounded by hungry pigs and secondly it is better and much safer to feed dry pig pellets.

Spread about on the ground, concentrated pig pellets also give the less dominant pigs a better chance of getting a fair feed while encouraging them to forage. When feeding food dry it is most important that pigs have constant access to fresh water. Lack of water can specifically affect the development of a young, growing pig. If watering is a problem then feed a 'wet' feed mix with a consistency of porridge.

Trough feeding and the '20 Minute Rule'

Feeding in a trough rather than on the ground has a distinct advantage when it comes to removing from a pen the feed the pig hasn't eaten. A very good rule to follow, whether feeding from a trough or on the ground, is the '20 Minute Rule'. If your pig has not eaten all its food within 20 minutes pick up the food and put it safely outside the pen – this is so much easier if it's in a trough.

If a pig does not clear its food within 20 minutes something could be wrong. Your pig could be showing the first signs of illness, so don't take a chance and leave the uneaten food down. If you do and your pig doesn't want to eat at the next feed time, you will not know whether this is because it is sick or because it has only just eaten what you left down. Pick the food up and put it outside the pen: if your pig was playing the 'I'll eat it when I want to' game it will cease when it realizes that the food is taken away if it's not eaten.

A pig that is not eating, or not showing a keen appetite, is usually a sick pig; the earlier you spot the signs of sickness the better. There may, of course, be another reason why it is not as keen on eating as normal. A pig in season that is 'brimming' may go off her food because 'sex takes priority over food'. You can find out by checking the vulva to see if it is swollen and pushing on her back.

Insight: The ice breaker

Frost was the curse of my life when I found that there is nothing worse than trying to unfreeze a garden hose or water pipe while surrounded by very thirsty pigs. I always make sure I have some barrels full of water near to my pigs all winter long just in case the pipes freeze up.

Dairy products and milk

It is illegal to feed your pig anything at all that has come from a recognized catering establishment and this includes eggs, milk and egg- or milk-based products. You must not feed eggs or egg products from any source whatsoever but milk from your own stock or from a neighbouring farmer who has surplus is perfectly acceptable.

Pigs do, of course, love milk and if you have a surplus of either cow's milk or goat's milk that has not come from a catering establishment they will thrive on it and the meat will be even more delicious than normal. But don't let the pigs get fat – all traditional and rare breeds have a tendency to put down fat very easily. They also love sugar beet shreds which satisfy their sweet tooth while also making them rest more. Once again don't overfeed despite the temptation. If you keep to no more than a third of their feed as sugar beet shreds you won't go far wrong.

If you mix pigs of differing sizes for any length of time the smaller pigs will always lose out at feed time and you will end up with some over-fat pigs plus small underweight ones also. Neither of these situations is good. Mixing pigs of differing sizes works as far as preventing fighting is concerned (the big ones see no threat and the smaller ones don't think it is worth it!) but watch feeding time and 'creep' feed the smaller ones where the larger ones cannot get at it.

Pigs are natural foragers and in an ideal situation like to eat little and often. Feeding your pigs morning and evening means that you can monitor their health and well-being twice a day. Control any urge you have to add that 'little bit extra' because although I know that it is a constant temptation with rare and minority traditional breeds because of their endearing nature, an overweight pig is no good to you or your pig.

If your pigs are outside (woodland and field environment) then during the summer months they may not need as much feed especially if the grazing is good and they have access to acorns and the like. Although you may be giving them less food you should still check your stock twice a day. You are sure to regret it if you don't.

Insight: Don't just feed twice a day

Feeding a pig is easy ... the hard part is remembering to vet it at the same time. I want healthy pigs so I vet them twice every day when I give them food.

Fruit and vegetables

Take advantage of the opportunities open to you by contacting your local greengrocer, market trader or supermarket and ask if you can have their waste fruit and vegetables. You will be amazed at the amount that is discarded whether as trimmings (cabbage leaves, etc.) or just fruit and vegetables that are past their best. There are some things that you should avoid feeding, especially anything citrus (oranges, lemons, etc.) because pigs are liable to get ulcers if they eat too much acidic food. You should also avoid parsnips because they can cause a pig to get mouth blisters and may contribute to infertility.

When feeding plums, cherries or even avocados you have no need to remove the stones as your pigs will get great delight in using their extremely powerful jaws to crack them open and digest the kernels.

Insight: Going nuts!

I have wasted more time than I care to tell you peeling bananas and de-stoning plums, avocados, cherries, etc. that I got as waste from my local supermarket, because I wanted to make life easier for my pigs. Don't do it. Pigs love the kernels, nuts and inside of any fruit and willingly eat the banana skins.

If you are feeding your pigs fresh fruit and vegetables they will be of greater benefit fed in between normal meals. Pigs love lettuce, tomatoes, cucumber, bananas (including the skin), apples, pears, plums and all soft fruit. Carrots, cabbage and potatoes are also good for them but you should boil potatoes before feeding them to pigs as they get more benefit from them. Strangely enough, pigs are not great lovers of mushrooms although they may eat them if cooked and the same applies to onions.

Barley meal is good for pigs but should not be the only food your pig gets as it does not contain enough vitamins, minerals and proteins.

Feeding procedures

Generally a pig should not be given more than 6 lb (2.7 kg) of dry food (nuts or meal) a day. When you collect your first piglets you should also buy at least one bag of the food they have been eating from the breeder. You can then keep feeding them the food they have been used to getting and over the course of a week gradually mix in your own feed so that you change their food over gradually. This will help prevent them getting stomach problems or even worse.

If your young, newly born piglets are kept inside and are not getting their nutrients and vitamins from the soil then you may have to give them an injection of iron.

Insight: For my good or their good?

When my feed suppliers said that they were dropping their feed bags from 25 kg to 20 kg so it would be easier to carry and meet with health and safety regulations, I was delighted. That is until I found from my next invoice that the feed price per bag had remained the same!

One exception to the maximum 6 lb (2.7 kg) rule is just before farrowing and when the sow is feeding piglets. Add 1/2–1 lb

(0.2–0.45 kg) of food a day per piglet born to the sow or gilt's normal ration after she has farrowed. Increase her feed by 2–3 lbs a day for the few days prior to farrowing. Lots of greens are also good at this time (cabbage, broccoli, etc.) as they help the sow produce milk.

Obviously make allowances for other food given (vegetables, potatoes, fruit, bread, etc.) and use your eyes to make sure that your pigs are never overfed. If you overfeed your pigs they have a tendency to put down fat. Some pigs, especially Kune-Kune 'pet' pigs, running on good pasture or in woodland need very little additional feeding during the summer months. However, if your pigs sink (start to lose weight) then increase the feed you are giving them as there is obviously not enough from the land to supplement what you are giving them.

When you approach your pigs at feeding time you want to hear a lot of noise because a noisy pig shouting for food is a healthy sign. It is the quiet, lethargic pig that you should be worried about and deserves closer inspection.

At times all pigs get boisterous and this normally happens at feeding time. When they are young, piglets may try to pull at your clothes or nibble at your Wellington boots (they love rubber!) when you go to feed them. This may seem fun as they are such delightful little animals but they will grow bigger! Being tugged at by a young piglet is one thing but when a fully grown pig does the same thing (especially if you have a bucket of food or water in your hand!) it is a different matter altogether. You could easily end up in the mud covered in pig feed much to the amusement and delight of passing neighbours.

Insight: Riding for a fall

Every time I think about the time a large Gloucestershire Old Spots put its head between my legs to get to the feed trough and nonchalantly flicked me into their mud wallow I smile, but it was far from funny at the time. It did teach me not to stand with my legs apart in front of the feed trough when feeding very hungry pigs who haven't yet learned how to say 'excuse me'.

Like all pigs they need to be trained otherwise they will develop bad habits such as the Wellington boot tugging game outlined above! A gentle but firm tap on the nose with your hand and they will soon learn what they are doing is wrong. If you watch a good mother with her piglets she does exactly the same thing by pushing them away with her snout if they are doing something she doesn't like. They may squeal but they are not hurt and soon learn that mother doesn't approve. Don't ever tell a pig keeper who has watched a sow bringing order to her piglets that discipline doesn't work! A pig is an intelligent animal and is easy to train when young which will help ensure you have years of happy pig keeping ahead of you.

Insight: Teaching our pigs good manners

My partner Carron (an equally passionate pig lover!) has taught a number of our pigs to sit politely while she feeds them apples. Visitors to our pig-keeping courses love to see that pigs aren't always messy eaters!

Concentrates, additives, genetically modified grain and growth promoters

Some of the concentrates, whether in the form of pig pellets, nuts, cobs or meal, supplied by animal feed companies may contain additives and growth promoters designed to assist in the quicker growth of the pig. This enables pig breeders to produce 'pork weight' and 'bacon weight' pigs more quickly, perhaps in half the time it takes normally for a free range naturally fed pig. Many concentrates also contain chemical additives that not only produce quicker weight gain (as a result of a higher percentage protein content) but are also aimed at protecting the growing pig against disease.

It is, of course, a matter of personal choice how you feed and fatten your pig (provided you stay within the law) but if you want to avoid all feeds containing excess additives and growth promoters then speak to your agricultural supplies merchant or local feed company. There are several very good feed companies that manufacture excellent 'natural' feed that will meet all your requirements and there are also many small mills that will make a mix especially for you. I believe that if you attempt to fatten at twice the normal rate then something has to give. You can't have quality and taste plus an unnatural speed of growth. Why does supermarket pork so often taste like cardboard? There are many pig feeds now available through your local agricultural merchants that do not contain genetically modified grain or other 'nasties'. Demand for such feed is on the increase and therefore you can be sure that if you tell your feed merchant or agricultural suppliers what you do not want your pig feed to contain then they will have no problem meeting your request.

Insight: Food good enough to eat

'Let me see you eat one of those pig pellets,' I once said to a feed specialist from a major feed company. He wouldn't, so I didn't buy that feed. If it contained 'nasties' that he wouldn't eat, then I wouldn't give it to my pigs. Because our current supplier makes feed without chemical additives, growth promoters and the like, they are always prepared to eat a few pellets whenever they visit. It may not be nice but it reassures me.

Most feed companies are concentrating their efforts on perfecting a satisfactory 'green' pig food that will have 18 per cent protein for weaners and growers or 16 per cent for sows, gilts and boars. There is even a complete 'Smallholder Range' of new 'green' feeds available on the market. Already the Vegetarian Society has given its approval to some of the feeds produced and although 'vegetarian approved' may appear an odd sign to put on animal feedstuff when the animal may end up as pork it simply means that it is a guarantee that your pigs are not eating any part of another animal.

After all the problems caused by foot-and-mouth as well as BSE and the horror stories that have come to the public's attention as a result I am sure that it is more important than ever to feed all livestock as naturally as possible. Even if the cost of these natural feeds is not always comparable with standard feeds containing growth promoters and additives, I am sure that the benefits outweigh any small extra cost. Your health, the health of your children, your friends and relatives must come first – not to mention the health of your livestock. If you want to produce pork you can be proud of and pork that tastes as it used to before the days of 'factory farming' when financial gain and commercialism took over from taste, quality and caring, then how you feed your pigs is of paramount importance.

'Never feed your pig'

I am ending this section with what you might think is a rather stupid declaration but as the next chapter deals with the health of your pig I thought that this was an ideal place to make such a prolific statement. I want you to keep this statement foremost in your mind for as long as you remain a pig keeper. Although you will realize that if you took this statement literally you would not be a pig keeper for long, at the start of the next chapter you will see why taking this mental approach to pig keeping will do more than almost anything else to keep your pigs well and happy.

10 THINGS TO REMEMBER

1   *Three is the magic number. Feed a fully-grown pig 3 lbs of pig nuts in the morning and 3 lbs in the evening.*

2   *Take the mental approach that you NEVER feed a pig. Instead, 'vet' them at least twice a day and at the same time give them some food.*

3   *It is illegal to feed your pig meat, meat products or anything that has been in contact with meat.*

4   *Constantly update yourself with the current rules and regulations concerning what your pig can and cannot have by reading the DEFRA website or by checking with your local Trading Standards Office (Animal Health) department.*

5   *The 20 Minute Rule is one you should always follow.*

6   *Make sure your pig gets fresh water **at least** once a day.*

7   *Waste or outdated fresh fruit and vegetables (very good for pigs) MUST NOT be stored or taken into your kitchen.*

8   *Outdoor pigs fed naturally will be ready for pork at between 26 and 30 weeks (summer and winter) and ready for bacon at between 10 and 12 months.*

9   *A pig who doesn't eat is usually a sick pig.*

10   *Sex takes priority over food. A pig that is 'brimming' (in season) may not eat as ravenously as she normally would.*

7

The health of your pig

In this chapter you will learn:
- *how to quickly spot a sick pig before it gets serious*
- *some of the more common complaints*
- *the legal side of managing sickness in pigs*
- *why 'never feeding a pig' is the best way to ensuring good health.*

In this chapter you will read a lot about some of the illnesses and health problems that pigs can succumb to. However, remember that it is highly unlikely that your pigs will encounter all or even any of these diseases or illnesses, but if they do you will be prepared. All too often, small-scale pig keepers feed their pigs with their eyes and ears closed to what is really happening. Feeding a pig is easy and as a result can be taken for granted. You may be under the impression that all you need to do is put pig feed in the trough before cleaning and filling the water bowl which is just a quick five-minute job. If this is what you think then you couldn't be more wrong and as a result you will end up with more sickness or ill health than should be the case.

Too many pig keepers, many of whom should know better, think that their twice daily visit to their pigs is simply to feed them but if you are a good pig keeper that is not the case. You should NEVER FEED A PIG because any fool can do that by simply throwing food down and filling a water bowl. For the sake of your pig and to become a good contented pig keeper you must take a totally different approach. Whenever you are taking food to your pigs you must mentally accept that you are not going to feed them but you are actually going to 'vet' them.

Of course while you are vetting them you will give them some food but your eyes, ears and mind should be vetting them all the time. If you vet them twice a day and at the same time give them some food, the longest your pig will be sick without treatment will be between feeds. If they were all well in the morning and yet when you vet them in the evening you notice something wrong then you know that they have only had the problem for a very short time. If you react to any illness or problem quickly then the chances of success are much better and recovery much quicker.

Insight: Happiness is pig shaped

'I can see your pigs are all fit and well,' said one of my pig keeping friends as he got out of his car. I hadn't seen him for months and he hadn't seen my pigs so I asked him how he knew. 'Don't you know that you can always tell how the pigs are by a pig keeper's face – he has a smile on his face that his pigs put there.' How true that is.

You must never forget how important you are to your vet. If you have any doubts about that then just think about how often your vet sees and vets your pigs and also what condition they are in when he/she sees them. Your vet sees and vets them very rarely and yet you see and vet them twice a day whenever you give the food. Your vet only ever sees and vets them when they are ill and you have asked for them to visit, but you see them every day in sickness and in health. You should be your vet's eyes and ears because you know more about your pig and its behaviour than your vet ever will.

When it comes to sickness and disease your veterinary surgeon must always come first for he or she is the expert.

The first thing you must do when you get your first pigs or piglets is to register with your local veterinary practice. If you have several in the area then I would recommend that you visit them all to establish which is best for you and your pigs. It is often difficult to find a good specialist pig vet because there are very few small-scale pig keepers about and therefore the demand to specialize is not there. You will probably be faced with using a general practice vet or a vet who concentrates on farm work, so when deciding which of your local vets will be right for you and your pig you will need to ask the following questions:

- ▶ *Do you deal with farm animals?*
- ▶ *Do you have anyone who specializes in pigs?*
- ▶ *Do you have many other pig customers?*
- ▶ *Do you deal with 'pet' pigs?*

The answers to all these questions combined with how near the practice is to where you are living, surgery times, how easy it is to get to and, of course, how well you get on with the receptionist and the vet will all play a part in the final decision you take regarding the vet you register with. Do not take what might seem the easy way out and do everything on the telephone because it is only by seeing the surgery and meeting those involved face to face that you will be able to get a true picture.

Not only is it correct and courteous to register with a suitable veterinary practice in your area but it also means that should your pigs ever need treatment you do not have to delay the process by having to ring round and find a suitable vet, then register, explain about the breed, and inform the vet what you are doing and how long you have had your animals. You will have done all this and will already have developed a relationship with the vet you choose even before your pigs are ill. This is a much better position for you to be in.

On your first registration visit, take a picture of your pigs, explain how old they are, how long you have had them (register in the first week they arrive or even before they arrive), why you are keeping them (pet, breeding or for meat), provide some background information about the breed (you must not expect your vet to know everything about every breed there is) and build a relationship with your vet. Even suggest that they can call in to see how and where you keep your pigs even before there is a need for an official 'I have a sick pig' visit. You may think that you are going to a lot of trouble when you don't even have a sick pig and may never need a vet's services but any effort you put in at an early stage will stand you in very good stead later on.

Although I hope that you do not experience any major health problems with your pigs it is almost inevitable that, as with all other animals, your pig will be unwell at some point during its lifetime. It is far less likely that you will have a sick pig if you are just keeping a few for the freezer. In this case you will only have your pigs for around 18 weeks and, provided you bought them from a reputable breeder who wormed them before you collected them, you should have strong, healthy pigs right through to the time they go to the abattoir. However, you must be prepared for the worst and if you are keeping a 'pet' pig for its entire life (possibly 15 years or so) or if you decide that breeding is the right way forward for you then you will inevitably have a sick pig at some time or another.

The one thing you must remember is that the longer you leave a sick pig without treatment then the less will be your chances of your pig making a satisfactory recovery. Pigs treated quickly recover equally quickly so you should not leave anything to chance and, when necessary, call on your veterinary surgeon as soon as you think something is wrong.

Your vet may need to visit to examine the sick pig but if you have done a good job of describing the symptoms it may be that your vet will be able to recommend a course of treatment over the telephone. This may involve collecting the recommended treatment from the surgery so you can administer it yourself. Whether your vet considers a visit necessary or not, the speed with which you contact the practice is of paramount importance to your pig and your wallet. The longer you leave it before you speak to your vet then the more likely it is that your pig will get worse. The worse the illness or infection becomes then the longer it will take your pig to recover and that means a longer treatment time, possibly more drugs and a much larger bill at the end of it all.

> **Insight: The call out charge**
>
> With the current average call out charge for a vets visit rising
> even as I write this book it didn't take me long to realize that
> learning the basics of examining, injecting and treating a sick
> pig was one of my first priorities.

If you learn to 'use your eyes' each time you visit your pigs you will
soon become very competent at spotting signs of sickness. If you
remember that you never just 'feed a pig' and instead 'vet it' twice
a day while giving it food, you will have happier, healthier pigs.
The following are signs to watch out for:

▶ *Not eating food or eating much less. This is a sure sign*
  something is not right. But beware the 'brimming' (in season)
  sow or gilt that may go off her food at this time because her
  sexuality takes priority over eating.
▶ *Drinking much less (or more) than is usual.*
▶ *Discharge from nose, eyes, mouth or vagina.*
▶ *Lameness and inability to walk properly.*
▶ *The colour and consistency of dung (becoming a 'dung expert'*
  is very important because the colour and consistency of the
  dung will tell your vet a lot).
▶ *The quantity and colour of urine.*
▶ *Generally lackadaisical and not 'bright eyed'.*
▶ *The rate of breathing and whether your pig is puffing, panting,*
  coughing or wheezing.
▶ *General condition of coat. Check for lice which often occur*
  on outdoor pigs. Does the skin look red and is your pig
  scratching a great deal (possibly mange)? Are there any red
  spots on your pig's stomach (possibly pig pox) or lumps on the
  side (possibly Erysipelas)?
▶ *Cold rather than warm extremities (i.e. base of tail and base of*
  ears) is often a sign of ill health.
▶ *Coughing and wheezing (which may be a sign of lung worms,*
  especially in young pigs).
▶ *Enlarged scrotum (possibly a testicular hernia).*
▶ *Lump on the navel (possibly an umbilical hernia).*

▶ *Is your pig's tail curly when it is walking (good sign)? A pig's tail will often go straight when it is standing still or eating but it should curl up tightly when it walks. This is not the case with Kune-Kunes, Potbellied and German Micro pigs.*

You will be surprised how quickly you can learn to 'use your eyes' every time you visit your pigs. It is always better to err on the side of caution even if a phone call to the vet simply puts your mind at ease. Pigs kept properly are not normally sickly animals but learning to observe your pig closely will not only help ensure good health but it will also teach you more about the animal itself. The pig is a fascinating and intelligent animal and 'pig watching' can become addictive – beware! If your pig appears ill or acts as if something is not right then take its temperature which should be around 102.5 °F degrees (39 °C). A thermometer is essential (two are even better!).

Insight: Getting close to your vet

One of my very best friends (he accompanied me to Austria/Hungary to bring the 'Curly Coated Mangalitzas' back) is a vet and over the years he taught me more about veterinary and pigs than anyone else. Although we now live many hundreds of miles apart he is still our vet because his knowledge travels with me. That's what you call a friend.

Worming

You should worm your piglets at about two months of age or when you first get them if you are acquiring weaners. You can ask the breeder to worm them before you take them home but if you do ask for this to be done try to arrange that they are treated in front of you so you can see it is actually done. All good breeders will do it automatically but I have known cases where new pig keepers have collected pigs they thought were wormed but clearly were not. You should then adopt a worming policy that ensures worming at least twice a year. Sows and gilts should be treated when they go to the boar and again about seven days before farrowing in order to prevent the placenta crossover of worms from the mother to the unborn piglets. If the piglets get worms in this way there is nothing you can do about it because they are already born with lung damage. They will be born normally and will show no outward signs but although they will suckle, apparently normally, they will be very lethargic and spend all the time sleeping. They will grow but will not play actively as piglets normally do and will die usually within the first ten weeks. If you are fattening 'for the freezer' your pig must not be taken for slaughter until the appropriate 'withdrawal time', as stated on the packaging, has elapsed.

A pig that is not wormed will not grow well no matter how much you feed it and 'lung worm' can easily kill younger weaner/grower piglets. It is a false economy not to worm your pigs on a regular basis for it is better to invest in wormer than to constantly be feeding your pigs without them gaining weight or pay a vet to cure something that you could have prevented happening in the first place.

You can administer wormer by injection, by using a 'pour-on' (a liquid wormer on the back of a pig in a similar manner used when administering flea repellent to a dog or cat), by liquid in the water, by powder in the food or by a 'drench' in the mouth. Choose the method you prefer to use and worm regularly. Some wormers also offer protection not only against all types of worms but also against lice and mange mites. From time to time, if you are keeping pigs longer term for breeding or as 'pets' you should change your wormer brand to ensure that your pigs do not build up any immunity. The choice of wormer is a personal matter but if in doubt take your vet's advice.

Enteritis vaccine

As a protection for your piglets, all your breeding sows and gilts should be injected with a piglet enteritis vaccine two weeks before farrowing. Gilts should also receive an additional injection six weeks prior to farrowing. This vaccine protects young piglets against E. coli infection and gives early immunity against a lot of diseases. It is important as it protects piglets during their first two weeks of life.

Erysipelas vaccine

You should vaccinate all your breeding stock against swine Erysipelas. It is a bacterial disease that is characterized by septicaemia and chronic lesions with the clinical signs of the septicaemia being high fever (temperature 105 °F) and very distinctive lumps or raised area on the back and or side. If your pig is not too hairy you will see them as 'diamond shaped'. Again, the first signs will be that your pig does not eat and just wants to lie down. If, on running your hands over its body you find raised areas or lumps then suspect Erysipelas and call your vet immediately. Penicillin will usually cure the problem if you act quickly enough but it may result in your breeding stock becoming infertile so it is much better to vaccinate.

Lice

Pigs often get lice. If your pig has lice they will easily be seen crawling around the back and neck of your pig with your attention having been drawn to their presence by the way your pig is constantly scratching and rubbing against anything that can stop the irritation. I think you would be wise to take preventative action by using a 'pour-on' solution or an injection but if you do get an outbreak of lice, a good covering with louse powder will solve the problem. Do not panic if you see lice on your pig because it is not a major problem. It is, however, possible that it may either be picked up from buildings or from other stock that is not protected. If lice are found, treat all your pigs and not just the one with lice. You may spot small dense areas of lice eggs around the neck and ears of your pig even before they have hatched so treat your pigs immediately rather than waiting until they start scratching and showing their discomfort.

Mange

Mange is another fairly common problems for pigs. It is identified by scurfy areas on the legs, particularly below the knee and hock. It may also be seen on the forehead, between the eyes to the tip of the nose and behind the ears. The area will be crusty and scurfy, sometimes looking like sunburn with apparently peeling skin. There may also be some quite severe biting of the ear by other pigs where the mange mite is causing irritation. You will not be able to see the mange mite with the naked eye in the same way that you can spot lice but your pig will show the same characteristic rubbing and scratching, often running the full length of a paddock trying to get away from the constant irritation.

As I believe that prevention is better than cure I have always injected against mange using the 'all in one' injection that covers lice, mange and worms. If you do get an outbreak of mange then a good covering with a suitable mange wash should cure the problem. Make sure you cover yourself well and use rubber gloves before administering the mange wash. The most effective and safest method of administration that I have ever used is by using a container usually used for killing greenfly on roses. Fill the spray with a mange wash solution and spray your pig thoroughly on the back, flanks and stomach, avoiding the eyes and then wearing rubber gloves rub thoroughly into the skin.

The real secret is, of course, in preventative treatment as well as cleanliness in your buildings. It is always good husbandry to thoroughly clean and disinfect all buildings regularly. If your pigs suffer from either mange or lice you must clean and disinfect the building as well as treat your pigs if you want to be sure of stopping another outbreak.

Plate 1 Kune-Kune.

Plate 2 Red Mangalitza.

Plate 3 Middle White.

Plate 4 Berkshire.

Plate 5 Large Black.

Plate 6 British Saddleback.

Plate 7 The Tamworth.

Plate 8 British Lop.

Plate 9 Oxford Sandy & Black.

Plate 10 Potbellied Pig.

Plate 11 Blond Mangalitzas.

Plate 12 Swallow Bellied Mangalitza.

Hernias

There are two types of hernia you may come across, both of which are reasonably common – the 'umbilical' hernia (a lump or swelling around the 'belly button' area of your pig) and the 'testicular' hernia characterized by enlarged testicles. If your piglet or weaner has a hernia it can be operated on but it is rarely necessary and is not something I would recommend. If the pig is intended for fattening, is not in any discomfort, is eating well and showing no other signs of illness then it is best left. The meat will not be affected and although a little unsightly it will not affect the pig. You may find it comes and goes, while in some cases it will disappear altogether before the piglet is even weaned. The trauma of an operation can cause the pig more stress than the hernia itself. A hernia is simply a rupture that may be caused by strain or by being trodden on by another pig.

In some cases, a hernia may be a congenital abnormality. If there are a number of piglets from the same litter born with hernias or that develop hernias as young weaners then you should seriously consider changing or culling the boar. If you have used artificial insemination or hired a boar then you should inform the insemination bank or the owner of the boar of your problem.

Tail biting

If your pigs show signs of tail biting (it usually only happens if the pigs are kept inside) then this is probably caused through boredom. Hang a piece of chain, an old rubber tyre or something similar from the roof of their pen or sty so they have something to play with. If you have done this and they also have a good amount of exercise but still show signs of tail biting, check for mites, etc. and call in your vet if in doubt. Outside pigs should not suffer this problem as they are generally contented and well occupied.

Farrowing fever

This condition can affect your pig after giving birth. If your sow or gilt is lethargic, does not eat or appears constipated (no signs of dung in or around her pen) then call your vet. Do not waste time or assume she is too exhausted after the farrowing process to be bothered about food because the reverse should be the case and she should be eating more not less. She could be suffering from farrowing fever and if caught early she should respond well to antibiotics which will bring her temperature down. If you can get her to eat again quite quickly then her milk will not dry up and she will be able to suckle her piglets successfully.

If the sow or gilt does not respond quickly you would be wise to prepare some milk in order to supplement what the piglets are getting from their mother. Proprietary baby milk from your chemist or supermarket is excellent mixed exactly as you would for your own baby. It has the advantage over specialist 'pig milk powder' of being easy to obtain and prepare, contains everything your piglets need and comes in much smaller, more manageable and less costly quantities. The piglets may or may not need it but it is better to err on the safe side. If your piglets are noisy and do not appear content or asleep in a heap, they are probably hungry and need milk. Hopefully, as is normally the case when farrowing fever is caught early, the mother will respond and all will be well but don't delay in taking action otherwise you could find yourself with a dead sow and lots of piglets to bottle feed by hand – every two hours!

Mastitis

Once again the first signs are usually indicated by the sow failing to eat her food. On feeling her mammary glands (udders) you will find the infected udder warm to the touch, lumpy or hard, painful to the touch and your sow will have a raised temperature. Mastitis often happens within two to three days of weaning and you should therefore watch her even more closely than you would normally do at this time. If your sow develops mastitis, speedy treatment is essential with a call to your vet being your starting point.

Lameness

Pigs that live outside, often on rough, hilly and stony ground can (and sometimes do) damage their clays (hoof/trotter). Whenever you see one of your pigs limping or finding it difficult to walk it may have damaged or split its clay so examine its foot and if you find a split or the trotter is tender to the touch your pig will need treatment and you will need to call in your vet. The vet will probably pare back the split clay, clean out the infection, treat the wound with a Terramycin (Oxytetracycline) antibiotic spray and administer a long-acting antibiotic. It may be necessary, if the weather is bad and the ground muddy, to keep your pig inside to stop any further infection and you should also spray the wound daily for three days as a further precaution.

Lameness in piglets and weaners could also be joint-ill caused by a 'bug' gaining entry through scratches or wounds caused by fighting. It usually takes two or three weeks to take effect. The bug goes to the joint and starts to multiply, causing painful inflammation around the knee joints and acute lameness – Pen and Strep will soon sort out the problem so call your vet and deal with your piglet as soon as possible.

Both damage to the clay and joint-ill will respond to treatment if dealt with quickly by a qualified vet. It may seem cheaper to deal with things yourself but in the long term it generally isn't the case unless you have gained sufficient experience and expertise. It's also not fair to your pig!

Lameness may also be caused by a strained muscle or ham. This often occurs when young piglets fight among each other to sort out a 'pecking order' and also among older pigs when mating takes place. Both the boar and sow are susceptible and rest, away from other pigs for a few days, will usually solve the problem but if it persists – call your vet.

Staggering

A pig staggering, apparently 'drunk' is not a pleasant sight for anyone, especially someone new to pig keeping. An apparently fit and well pig one day can develop 'drunk' symptoms overnight and the causes can be as follows:

1 **Bowel oedema.** *In this case the staggering is accompanied by swollen eyelids. This usually occurs in freshly weaned pigs and can result in sudden death of the best pigs in the litter. It is caused by an E. coli bug although stress, change of environment and change of food are thought to be contributory factors. (Always wean piglets by taking the mother away to a different area, not the piglets. This will help prevent the combined stress of losing their mother and having to get used to a new environment.) Because the brain is affected it is rarely possible to save the piglet. Seek your vet's advice.*

2 **Sunstroke.** *If it has been very bright and your pigs are exposed to the sun's rays on a hot day then they could develop heatstroke. This will cause your pigs to shiver and stagger. Pigs that are burnt by the sun's rays don't want to move and if you try to force them they will wobble and collapse, going off their back legs. Treat by moving to a cool, dark area and if severe heatstroke is diagnosed, drench or spray with water to bring the temperature down. Soaking a towel in cold water and placing it over your pig will also help. Keep your pig inside and out of the sun's rays but with plenty of good, clean bedding until everything returns to normal. Antihistamine injections can be given against shock. Always provide good shelter from the sun and of course a 'wallow hole' as the mud will help protect them from the sun.*

3   **Salt poisoning.** *Symptoms include staggering, running backwards, quivering or collapsing. Your pig may have a strained expression and start champing of the jaws and frothing at the mouth. Caused more by water deficiency rather than feeding too much salt. Water is needed to dilute the salt content of food so treatment is not difficult and the results are easy to see. Increase the water supply (clean water) immediately. Make sure that, if you keep sheep, your pigs do not have access to the salt licks.*

4   **Actually drunk.** *You may have fed your pig too much old fruit that was in the process of fermentation. If you get a good supply of old fruit from your greengrocer or supermarket beware of overfeeding. Pigs can and do get drunk very easily. If your pigs are grazing an orchard make sure that you pick up some of the fruit if there is an excess of 'fallers'. Eating too many that are overripe will cause them to become very drunk. Older pigs don't take long to learn that if they scratch themselves against a tree they are greeted with the delights of fruit galore.*

Insight: One over the eight

I once found a pig in our orchard lying under an apple tree not moving a muscle having apparently had a heart attack or similar. It was in fact just 'drunk as a skunk' having eaten a whole basket of overripe fallers that I had left there by mistake. Pigs can get drunk very easily on too much overripe fruit.

The law and animal health

You are required by DEFRA to keep an Animal Medicine Record Book which you can probably get from your local Ministry of Agriculture office or from your local Trading Standards Office (Animal Health) and if not they should know where you can get one. If you can't get a 'free' one then you can easily draw one up in a notebook. You need the following headings and columns:

- ▶ *Date medicine purchased.*
- ▶ *Identity of animal treated.*
- ▶ *Date treatment finished.*
- ▶ *Length of withdrawal period.*
- ▶ *Batch number of the drug used.*
- ▶ *Earliest date after treatment the animal may enter into the food chain.*
- ▶ *Name of person or vet administering medicine.*
- ▶ *Name of medicine.*
- ▶ *Date treatment started.*
- ▶ *Quantity used.*

You are also required under the Animal Health Act to get a licence for the movement of any pig (other than for slaughter or piglet to the vet) from your premises.

These forms can be obtained from your local ministry officer and you can now make out a Movement licence for yourself on the day of the movement. You need a separate form for movement for the purposes of breeding. One copy is given to the person who is moving the pigs and the other copy is sent to your local ministry officer. If in doubt go to your local fat stock market and speak to your local inspector – inspectors are usually very helpful.

You are also required to keep an up-to-date Movement record of all your animals. The appropriate Movement book can be obtained from your local Trading Standards Office (Animal Health).

DEFRA requirements

DEFRA requirements for identifying pigs is all about good health and traceability. Before you have any pigs on your premises you must get a CPH (County Parish Holding) number. To get this you simply telephone the appropriate Rural Payments Agency (RPA) number and inform them that you require a CPH number because you will be keeping pigs. You will find the correct RPA telephone number by visiting the DEFRA website or by contacting your local DEFRA office. You will also need a herd number for your pigs and this will usually be issued at the time you get your pigs and sent in your first Movement licence.

You are required to tag (metal ear tag) or 'slap mark' (a form of tattoo on the pig's flank or shoulder) all pigs that you send to the abattoir or all pigs over 12 months of age that are moved off your premises. The metal tag or the 'slap mark' contains your UK herd number and assists traceability of all stock, dead or alive. I personally prefer the metal tags as they cause virtually no stress or discomfort to the pigs. Any pigs, over 12 months of age, entering your farm must have the UK herd number of the farm they originally came from. These identification details along with your CPH number must appear on the Movement licence whenever you receive or dispatch pigs over 12 months of age on and off your premises.

Pigs under 12 months of age (other than those going to the abattoir) need not have a 'slap mark' or tag and can be moved on and off your premises provided they have a simple colour identification on their back or shoulder while being moved.

This form of identification has nothing to do with individual breed identification as required by the British Pig Association. This is a separate issue for the registration of individual rare breeds. But it is very important for the continued good health of all rare breeds that identification is clear and that in the event of a serious outbreak of disease (such as foot-and-mouth) all our pigs can be traced to protect the healthy ones.

Insight: Having a quiet word with them

Talking to my pigs may not necessarily make any difference to their health or cure any illness they might have but it certainly makes me feel better.

General advice

The more you talk to your pigs and the more you handle them then the easier it will be for you to treat them if they do become ill. Your vet will also bless you if you have pigs that are friendly and manageable. Most traditional 'free range' pigs that are well handled and used to being treated kindly are docile, intelligent and very affectionate animals. Treated correctly, a pig will become a very trusting friend as it has nothing to fear, it has its freedom and has no reason to be nasty.

Insight: Eye to eye

I look into my pig's eyes and think 'now there's a bright eyed healthy pig' and she looks back into my eyes and probably thinks 'stop staring at me and get that feed in my trough'!

Do, however, beware during farrowing and where young piglets are concerned. Even the most docile of pigs can be ferocious (and extremely dangerous!) if she feels her piglets are being hurt. If a piglet squeals put it down immediately and always have something to protect yourself with if you are in a pen or shed alone with a mother and her piglets. The boards that are used for guiding and moving pigs were originally intended as 'protectors' so make sure you have one. Take your time and have patience when handling young piglets and they will soon get to know you and not squeal for help when you pick them up. The mother will also learn that you are not going to hurt them and will not react badly, but don't rush things.

Good diet is important for any animal and the better you feed your pig then the stronger it will be. It is important to keep your pigs in good condition if they are to be able to cope well with any illness. Feed and water both regularly and well (but don't overfeed!). Never take the soft option and feed two helpings in one or worse still two days in one. Give your pigs access to the soil wherever possible for they get a lot of their natural minerals from the earth. If you can't run them outside (especially during the winter) then toss a turf of soil in to them on a regular basis and they will love you for it.

Cleanliness is equally important when it comes to the health of your pigs. Pigs are clean animals. If you see a pig lying in its own dung or in a messy sty – it is the pig man who is to blame – not the pig! Clean bedding, warmth and protection against draughts are a priority for your pig if it is to remain in good condition. Clean out your pig's home regularly for it not only helps prevent disease building up but it also gives you a 'happy' pig.

Insight: Sign of being unwell

When my partner Carron says to me, 'Are you not feeling very well?' I know that she really means I am being 'grumpy'. On the other hand, if my pigs are 'grumpy' I know that they are not very well.

An outdoor pig can cope admirably with all weathers if it has protection against wind and shade from the sun. A pig is the only other animal apart from humans that suffers from sunstroke and heatstroke. Your pig will love you forever if you provide the facility for a 'mud wallow' when outside and in the hot summer months the healthiest pig is often the muddiest. If you pour water on the ground when it is hot then your pig will do the rest. If your pigs are constantly tipping up their water bowl and trying to lie in it they are trying to tell you something – they need water on the ground as well as in the bowl!

The expert in pig diseases and ailments is your local vet so, if in doubt, don't hesitate to call the surgery. Your vet probably won't have many pig keepers on his or her books and perhaps has never treated anything other than a commercial 'pink pig' but he/she is the professional nonetheless and will be able to give you lots of help and advice.

Insight: Sickness benefit

Every pig farmer, me included, knows that there is no disability payment or sickness benefit for pigs so it seems obvious to me that they are not ever just pretending to be sick.

What to do if you have a sick pig

Firstly, if possible, separate the sick pig from the group and bring it nearer to the house in a clean, warm shed. By doing this you:

1 *help prevent the spread of any contagious disease;*
2 *help relieve the sick pig from further stress as a result of being bullied and bothered by the other pigs while it is unwell;*
3 *provide a warm, secure environment for the sick pig which will improve its chances of making a quick, successful recovery;*
4 *make life easier for yourself as you can more readily observe the symptoms, progress or lack of it and the pig's general well-being; and ...*
5 *make life easier for the vet to observe and deal with the patient. Not many vets delight in dealing with pigs in a muddy, rain soaked field when a warm well-located shed is their ideal environment.*

Secondly, note all the symptoms and any reaction to treatment so you can inform your vet accordingly. It is information like this that your vet will need to make a correct diagnosis.

Thirdly, ask your vet the simple question – what is wrong and what has caused it?

This is the only way you will learn and prevent similar problems in the future. Note down what your vet says is wrong with your pig and then read as much as possible about the disease. This is good practice and will help you learn more about pig keeping for the future.

Most important of all make a note of what drugs your vet has used including dosage, batch number and how many days it is to be administered. This is needed for your Medicine book but it will also help you to learn what drugs to use in similar circumstances. Don't forget the batch number.

ONE FINAL QUESTION

Have you got a thermometer to take your pig's temperature (ideally around 102.5 °F/39 °C) – if not why not?

If the worst happens

When keeping pigs you must never ever forget that although they will bring you great happiness there are other sides to the equation:

▶ *If you are keeping pigs to fatten for the table you will have prepared yourself emotionally for their ultimate end by knowing that their time with you has been better than anything else they could have experienced in a commercial unit. You also know that the only way to save our rare breed pigs is to find a market for them.*

▶ *If you are keeping a 'pet' pig with an expected life span of 10 to 15 years or if you have a favourite breeding sow that eventually gets too old to breed from then you will need to accept that the sorrow of eventually accepting that nothing lasts forever is far outweighed by the happiness and pleasure your pig has given you over the years. I always feel that you owe it to any pig that has given you so much pleasure to continue keeping pigs as a 'thank you' to what could be a 'long lost friend'.*

Insight: Sad but it happens

There are times when you have to deal with death. When I first started keeping pigs it was permitted to bury piglets born dead or, if you had a suitable digger that was able to dig a large enough hole, even older pigs at the end of their lives or pigs that had died suddenly (heart attack or twisted gut) could be buried in your field. This is now illegal and dead stock must be taken to a properly licensed place of disposal. Contact your local Trading Standards Office (Animal Health) for further information if you should ever need to dispose of dead stock.

Sometimes despite all your efforts combined with the skill and knowledge of your vet it is not possible to stop the inevitable happening. It may be the result of an incurable ailment or even just old age as was the case recently with 'Damson' one of my favourite pigs. She was a much loved Oxford Sandy & Black Pig who, over the years, had produced many marvellous litters of piglets. She was a great mother but an even greater friend so when she died I wrote this as a tribute:

Just the two of us
  (My final moments with 'Damson' my favourite pig)

> Counting down the moments
> Letting thoughts rush by
> Always asking questions
> I seek the answers – Why?
> Reaching for solutions
> Even as I cry
> Another day
> No not that way
> Don't go away
> Vainly I have fought the fight
> Often without fear
> Now I know the time has come
> A final word to hear
> Could that word be 'Goodbye' a
> Long lasting friendship gone
> A final stroke, a fond farewell
> I know life must move on
> Remembering the good times
> Eventually we won
> Another day
> No not that way
> Don't go away
> 'Tony wants to tell you that
> Over weeks and years'

Now I whisper gently
As I stroke her ears
'Countless times you've given me
Love beyond comparing'
As gently now I just kiss your nose
I really know you're caring
'Rest well dear Damson as
Even now the final hours we're sharing'

Another day
No not that way
Don't go away
Very few will ever know
Or understand my grief
But now that we have spoken
There's a feeling of relief
Keeping pigs brings happiness
Achieving something new
Of finding something special
Something new to do
It does much more than that
By becoming part of you
Another day
So now I say
You're on your way
I will continue keeping pigs
Right to my very end
Because you brought me happiness
As a loved and trusted friend.

Tony York – 21.09.2009

No matter what happens in your own private world of small-scale
pig keeping or in your life generally there is no doubt in my mind
that as a small-scale pig keeper you will make an awful lot of four-
legged friends.

Insight: Life or death

'What happens,' I recently asked a new pig keeper, 'when a piglet is born not breathing and you pick it up, warm it, clear its throat, blow gently in its mouth and then tickle its nose with a piece of straw until it coughs, splutters and comes to life?' 'I don't know,' was the rather puzzled response. 'Very simple,' I responded. 'You cry. I've done it many times.' That's what saving a young life does to you.

10 THINGS TO REMEMBER

1 *Buy a thermometer before you get your pig.*

2 *Register with a local veterinary practice as soon as you get your pig. DO NOT wait until it is sick.*

3 *Do not procrastinate. If your pig is showing signs of not being well then act quickly. Sooner treated – sooner better.*

4 *Worm when you get your pig then regularly every six months. Worm your sow when she goes to the boar and again about 10 to 14 days before she farrows.*

5 *Always enter any treatment you give your pig in your Medicine book.*

6 *If a pig dies for any reason (old age or ill health) it cannot be buried on your land. It must be taken to a properly licensed place of disposal and incinerated.*

7 *You are as important as your vet. You personally 'vet' (not feed!) your pig twice a day while giving it food. Your vet only sees them rarely and then only when they are ill.*

8 *If they are not running outside, give newly born piglets soil to eat (a turf of grass and soil in their pen daily for a week) because they need the iron. Alternatively, give them an iron injection before they are a week old.*

9 *Fresh water is just as important as food.*

10 *Not only must you look after your pigs but you must also look after yourself and the family. Practise good hygiene all the time and make sure that anyone who goes in with the pigs always washes their hands. Use antibacterial soap or hand wipes after handling your pigs especially when children are involved. Preventing E.coli is about cleanliness and good hygiene.*

8

Breeding from your own stock

In this chapter you will learn:
- *the first steps towards breeding pigs*
- *when best to start breeding*
- *about farrowing and raising piglets*
- *the secrets of mastering artificial insemination.*

If, like most first-time pig keepers, you fall in love with the rare or minority breed you have chosen then you may start thinking about breeding your own stock. This means you will be involved in many more things to do with your pigs that are new to you. You should start by asking yourself the following questions:

▶ *Why do I really want to do it?*
▶ *How do I start?*
▶ *Where should I do it?*
▶ *When should I do it?*
▶ *What do I do?*

Insight: A wonderful experience

'Is anyone thinking of breeding?' I asked at one of our recent pig keeping courses. 'Yes,' came the instant response from one couple, 'but only with pigs!' I certainly prefer pigs to a lot of people I know and this couple seemed to prefer piglets to children.

Why do you want to become a breeder?

Before you start down the road of breeding pigs of any breed you must first think about why you want to breed pigs and what you are going to do with your piglets. All too often I have watched people who have kept pigs successfully as pets or as a means of producing a 'mouth-watering eating experience' for themselves, friends and family who decide to become breeders and fail. In fact, they have failed so miserably that they have never kept pigs again. To avoid this disaster think carefully before you make the move.

It is at this stage, even before you start breeding, that you need to plan your sales and marketing policies. If you are thinking of breeding just to enjoy a few extra delightful little piglets running around then you are in for a shock. Piglets grow up very quickly and the more they grow, the more they eat. You must not overlook the fact that the larger they get the more space they need. If you do not know what you are going to do with all those piglets before you start (one sow can produce 20 or more piglets a year), then you could find yourself regretting the move into breeding.

It is much better to have your sales and marketing strategy in place at an early stage rather than waiting until they are eating you out of house and home before pushing the panic button. It is sensible to have created a market for your piglets as early as possible and if they are born with a 'sold' sign on them then so much the better. A piglet sold earlier produces a smile on a pig breeder's face rather than the scowl produced by waiting until they are costing more to keep than you can afford. You certainly don't want to sell your stock at ridiculously low prices just to get rid of them because if you have to do this then you are unlikely ever to breed pigs again. This situation is not good for you or traditional rare breeds, especially if the original reason for breeding was to help save our rare breed pigs. It is right and sensible to aim to more than just cover your costs for in this way you will continue to breed and contribute much more towards saving our rare and traditional breeds from extinction.

How should you start?

Start by planning what you are going to do with 20 or more piglets over the course of a year.

Firstly, unless you are simply breeding 'pet' pigs, you can allow one for fattening for yourselves if you want your freezer full of good pork and delicious bacon. At the same time take into account the number of advance orders for pork and bacon you have taken already. If you have bought in weaners previously to fatten and this is your first venture into breeding then you will have a number of 'regular' customers already, such as friends, neighbours and relatives.

Secondly, you should aim to sell all your newly born piglets (except those you are keeping to fulfil advance pork orders) as weaners by the time they are eight weeks old. Some will go for breeding, some as 'pets' or garden 'rotovators' and others to be fattened by the purchasers themselves.

Thirdly, remember to take advance meat orders even before the piglets are born. If you know the whole and half pigs you have sold to regular customers, friends, neighbours and acquaintances well in advance (many are repeat regular orders) you can plan accordingly. All your customers should be aware that many of your litter are going to be sold as weaners and that you will only fatten piglets for pork if they ordered in advance. Sometimes people have to accept disappointment and often these people end up buying weaners to fatten themselves. It is less work for you if you can create a market for your piglets rather than keeping them for six months to produce pork. If you are breeding 'pet' pigs then it is even more important to make sure you have customers as early after farrowing as possible. Advertise two weeks before the piglets are born to ensure that when your prospective buyers visit they see adorable piglets rather than one adult pig.

If you start letting everyone know you will have real pork available shortly then you will be surprised how much interest it will generate even many months before they are ready. For considerably less than somebody would pay for individual joints anyone can have a really top-quality half pig in their freezer, tasting sweet and succulent like pork really should, free range and traditionally reared. If you are keeping one yourself and have enquiries for five more from family and friends then you will have accounted for more than half your litter even before they are born or you place a single advertisement. If your gilt has a litter of ten you should be very pleased – and already six are accounted for!

You may, of course, want to fatten all your stock for selling as pork or bacon but beware of putting all your piglets in one basket. I think it is better to leave all your options open so you can sell weaners first and then fatten the remainder to sell 'direct' as specialist pork. Whatever decisions you make don't wait until the piglets are born – that is the way to disaster, panic and loss-making breeding.

Where?

Be sensible and give your gilt the attention she deserves during her farrowing by putting her in a warm, draught-free ark or building as near to the house as possible. As a new pig breeder you will want to make life as easy as possible for yourself so being near the house is a distinct advantage.

You can safely run your 'in-pig' sow or gilt with the boar during the first two to three months after service but you must separate any 'in-pig' female (from both the boar and any other females) about three weeks before farrowing. This gives the pig time to prepare and settle herself down for farrowing in her own 'maternity wing'. If you do not do this then you could suffer heavy losses as a result of the 'in-pig' sow or gilt getting knocked about during the later stages of pregnancy.

Letting your pig farrow in the same ark or area as other pigs is not an option even if she is with her best friend. You could suffer piglet losses if other pigs in the same area were to lie on the young piglets and crush them. The mother may also be distracted by the other pigs and be less interested in her new family during the first very important two to three weeks. Fighting among adults (pushing each other at feed time) or even simple bullying (sorting out a pecking order) can also occur, which may also result in losses. The rule should always be to separate early and farrow separately.

Insight: A world record

One of our ex-course members telephoned one day to inform us that one of her two Large Black sows who were due to farrow had given birth to 21 piglets ... possibly a world record for the breed. It was only the following day we learned that BOTH sows had farrowed at the same time the night before and the piglets had joined forces without the breeder realizing. The lesson – always farrow in separate pig arks. Ten piglets live per litter is a good average.

Try to keep your 'in-pig' sow or gilt on comparatively flat land during the latter part of her pregnancy as over-exertion and strain can result in some, or all, of the litter being born dead.

When?

If you buy weaners and start your plans for breeding the slow, gentle but more sensible way then they will be ready for the boar from about nine months old. Although some breeders put their young gilts to the boar at a much earlier age I believe that leaving it just that little bit longer results in more mature, stronger mothers that are better able to cope with farrowing.

Some breeders do not put their gilts to the boar for the first time until they are at least ten months old. Some people would say they waste a month or two but they always have good litters from their gilts and prefer it that way. Their gilts grow into good mothers and big strong sows.

You could, of course, acquire an 'in-pig' sow or gilt from a reputable breeder and in this way save a considerable amount of time. However, a word of caution – when buying an 'in-pig' sow or gilt you are better (as with buying all your pigs) to actually visit the breeder and look at all the stock before you decide. Buying from a market or sale has many disadvantages, not least being that you cannot see the surroundings in which the animal has been reared and how the stock is treated.

How close to farrowing the sow or gilt is plays a large part in the price you will have to pay. The closer to farrowing then the higher the price – this is known as 'close to profit'. A sow producing you ten piglets that you can sell at eight weeks old is obviously worth more than a sow who is not yet 'in pig'.

You should also expect to pay more for a registered 'in-pig' rare breed sow or gilt. The price you pay will also depend on age, markings and how well she conforms to the breed standards, background, birth lineage and history of previous litters.

Insight: Magic threes

Gestation period 3 months, 3 weeks, 3 days, 3 hours, 3 minutes, 3 glasses of wine. Brimming (in season) 3 days ... between brimming 3 weeks. Artificially inseminate 3 times ... feed a fully grown pig 3 pounds of pig nuts twice a day. To this list my partner Carron has recently added the following for all lady pig keepers: 3 handbags and 3 pairs of shoes every 3 months, but I have never seen this in any other pig keepers' handbook!

Remember three months, three weeks, three days, three hours, three minutes and three cups of tea (112–15 days) is the approximate gestation period. You must plan when you want your piglets to arrive to suit your own needs, by answering such questions as:

▶ *Am I going to be on holiday when the pig farrows?*
▶ *When do I need more pork for the freezer?*
▶ *Is the accommodation I have warm enough for farrowing in the snow?*

> ▶ *When is the best time for me to sell any weaners I might breed?*
> ▶ *Should I plan to be away from work when farrowing takes place?*
> ▶ *Would it be nice to involve the children when they are on school holidays?*
> ▶ *Have I got other important commitments to consider?*
> ▶ *When can I best afford to be feeding extra pigs?*

You must decide when the best time is for you – everyone has different priorities.

Insight: When it's quiet and dark

I promise you faithfully that sows do not always farrow and have their piglets at 3 a.m. in the morning when it is very cold, dark, raining or snowing. But after keeping pigs for over 30 years, it just seems like they do!

Your gilt or sow will come into season (start brimming) once every three weeks (21 days). She will be 'on heat' for about 50–60 hours during the summer and about 48 hours or less during the winter and will not usually be ready for the boar until 12 hours after she first comes on heat. Mating that takes place before the 12 hours has elapsed may result in the ovulation being missed. It is therefore day two of the brimming period that is the best day to take your pig to the boar.

Watching for signs of your gilt or sow coming into season is very important if you are taking your pig to the boar. It is less important if you have your own boar (or are borrowing one) to run with your pigs but obviously you still want to know the date when your pig was served.

Signs to look for are:

1 *Reddening and swelling of the vulva. This usually lasts for about three days and is the first sign of approaching heat.*
2 *Standing still under pressure and weight on her back – a very clear indication indeed. If she stands very still when pressure is applied then she is ready for the boar – if she moves away*

when you push on her back then she is unlikely to stand for the boar and is probably not ready. Once she passes this test then she should be put to the boar immediately. If you rub her sides gently you will encourage her to stand motionless if the time is right. You will see the boar nudge and rub her side before he mounts and you are just mimicking this action.

3 *Moistening of the vulva occurs at the same time as the swelling and can easily be determined by touch when she is in heat.*

4 *Change of temperament and behaviour. Quite often there is a reversal of behaviour when the quiet gilt becomes restless or the less easy to handle gilt becomes docile. She may go off her food and have a reduced appetite while she may also do more grunting than usual (probably trying to tell you that she would like a boar, please!). You may also find she tries to get out of her paddock, lifts gates, pushes against fencing or paces up and down a perimeter fence in the expectation that a boar will be made available.*

5 *If you are running the boar with your female stock on a permanent basis then you may not actually see mating take place. In this case look for the tell-tale signs of brimming then look carefully twice a day for mud markings on the sows back. Mud marks usually appear more obviously around the top and side of the hams where the boar has been gripping her with his front legs. In wet, muddy weather this will be very obvious indeed. In dry weather you may also see signs of scratching along the sides of your sow or gilt where the boar has been gripping with his trotters.*

You should check for signs of brimming at least twice a day. If you have more than one sow or gilt in together you will undoubtedly find they start brimming about the same time.

You can either choose to take your pig to the boar (making sure your pig is brimming on that day) or you can borrow a boar and run him with your pig for 24 and 44 days to ensure you get the right day. By keeping the boar long enough to cover your pig coming into season twice (21–42 days) you help protect against your sow failing to conceive the first time and having to have the boar out again. Whatever you may decide you must make absolutely sure that you contact the owner of the boar well in advance so you know that everything is prepared and that a suitable boar is available.

Pure bred boars can be very much in demand and you may, of course, decide to buy a boar of your own but you must consider the cost implications, especially if you only have a few pigs. Working boars need to be fed more than sows to keep them in good condition and will, of course, need separate housing from the sows or gilts for some of the time. If you keep them together all the time you may find that you cannot get your sows 'in-pig' because familiarity breeds contempt and you may have to separate your boar and then try again six weeks later. If you only have a few sows then he may also get bored and restless with the added disadvantage that you will not want to use your boar on any of his offspring that you may keep for fear of in-breeding.

The cost of hiring a boar can vary from breeder to breeder but expect to pay a straight stud fee for each sow with reductions for more than one sow or gilt. Some breeders may even be prepared to accept payment in kind by perhaps accepting first choice of the resulting litter.

Artificial insemination

Whether to own a boar, whether to hire a boar or whether to use artificial insemination (AI) is a dilemma that many breeders face.

Insight: A fantastic achievement

The first time I used artificial insemination in 1970, the success rate was phenomenal because the sow had piglets just eight weeks after insemination instead of 16 weeks later – half the time! What a fool – I was so inexperienced I had given artificial insemination to a sow that had already been served by the boar eight weeks earlier! They all looked the same to me!

Since the disastrous 'Foot and mouth' (F&M) outbreak in 2001 the movement restrictions on animals have been tightened up and fewer people than ever are hiring out the services of their boar. You may find that finding anyone who has a boar that is the same breed as your sow/gilt and is unrelated is very difficult. On top of that you have to find someone in that position who is prepared to hire you their boar. It is by no means impossible but it is difficult – sometimes VERY difficult.

Your next option is to own your own boar but, as you have already read, if you only have a few pigs it is neither practical nor cost effective so you need to seriously consider artificial insemination.

Many people new or comparatively new to keeping pigs often shy away from the idea of artificial insemination, believing it is difficult, complicated or quite simply 'something they could never do' when in reality it is just a matter of following a few simple rules.

Your AI kit will come to you in a discreet polypropylene (or similar) container by first class post and will consist of the following:

▶ *Three sealed plastic bottles containing the appropriate semen.*
▶ *Three sets of applicators each in two parts. As the penis of a boar is like a corkscrew one end of the applicator is fashioned in this way.*
▶ *A leaflet giving you help, advice and instruction on how to administer the semen.*

Your success depends on not deviating from the following procedure:

▶ *First and foremost get the 'brimming' cycle (when she comes into season) exactly right. Check her twice daily for signs of brimming, looking for all the signs listed earlier. When she comes into season make a note in your diary and exactly 21 days later check again. If she is 'brimming' again when you check then 'bingo' you have got her brimming cycle right.*

- Knowing that she will come into season (brim) every 21 days, you can now order your semen and AI kit by telephone about seven days before the due date. Currently all rare breed semen comes from Ireland (see 'Taking it further – Suppliers of pig equipment') and when you give the date your sow or gilt is going to come into season the suppliers will post it to you at the appropriate time to ensure it arrives with you on or just before the appropriate starting day (she will 'brim' for three days).

- Remember that your pig is in season for three days and that of these three days she is most fertile during the middle 24 hours. This is the time when the semen is likely to be most effective and when she is most likely to conceive. Bearing this in mind use your three bottles of semen and applicators wisely. Administer the first bottle towards the end of day one, the second bottle at the start of day two and the third bottle late day two.

- If your sow or gilt is brimming she will 'stand' as if waiting for the boar to serve her and if you push hard on her back she will think that the boar is going to serve her and will stand motionless. If you want to put her in a 'crush' to restrain her or put her behind a hurdle, then do so but if she is truly in season then this is unlikely to be necessary although it may give you confidence the first time you do it.

- Before you insert the applicator you will need to lubricate it. DO NOT USE SOAP. Soap kills the semen so use a proper lubricant such as KY Jelly.

- Insert the applicator into the vulva (corkscrew end first) by turning anti-clockwise. When you remove the applicator turn clockwise.

- Insert the applicator pointed slightly upwards as you turn anti-clockwise because if pointed downwards you will end up in the bladder.

- When you feel the applicator 'lock', cut off the tip of the plastic bottle containing the semen (the semen should be kept warm in your pocket and should not have been put in the fridge) and attach to the end of the applicator. DO NOT squeeze the plastic bottle to speed up the process of the semen *entering the pig as this will welcome failure. Let gravity and nature take its course by holding it up in the air and allow the sow/gilt to flex her muscles in order to remove the semen naturally.*

▶ *When all the semen has entered the sow/gilt, take off the plastic bottle that contained the semen and remove the applicator by turning the applicator clockwise. Do not keep these as they serve no useful purpose and disposing of them will prevent you mixing old and new applicators. Use and destroy the other two new applicators that you still have at the appropriate times.*

▶ *Follow the same process each time. Use all three semen bottles and applicators over the three days she is 'brimming'.*

▶ *You should clearly mark the dates when you carried out artificial insemination and in exactly 21 days check if she is 'brimming' and if she doesn't come into season again buy some of that red or white medicine I have to take from time to time (it might come from Chile, Australia, France or Italy!) and celebrate because your pig is now 'in pig' (pregnant) and you are a very clever little new pig keeper and have taken another step towards helping save our Rare and Traditional pig breeds – well done!*

How?

Your gilt will know how to farrow but you need to make life as easy as possible for her. Pigs do have a tendency to roll down on their piglets and fatalities are not uncommon. If this happens you are by no means alone but if you practise good husbandry then the losses will be minimal (one or at the most two per litter should be the maximum). A farrowing rail or a sheep hurdle that the young piglets can get behind will help. A heat lamp is not absolutely necessary even in the winter but it does attract the piglets to safety if it is in an area where the mother can't get at. Instead of seeking warmth close to the mother (risking having her lie on them as she moves or gets up) they are able to congregate under a heat lamp.

Farrowing is not usually a long process, lasting possibly two or three hours but there are exceptions to the rule and an 'all night sitting' is not unknown. If you have handled your pig well during the run up to farrowing she will not mind you being present. Nature is a wonderful thing but a little of the right help never goes amiss. Being present is also good for you as a pig keeper because you will understand even more fully the wonderful natural instincts of both mother and piglets. Always have your pig board with you in the pen just in case your sow becomes upset at you being there. If you have bonded with her during the final weeks of pregnancy then it is unlikely that she will mind your presence but if she does, then leave her alone to settle down and cope with the birth herself rather than distress her. You can, and should, watch discreetly from time to time to make sure all is going well. Watching from an adjoining pen or over the door is fine but be quiet and don't have all your children, family and friends joining you. Would your mother have wanted half the neighbourhood watching while you were being born?

Insight: My strangest litter

I was sitting in a cold pig ark with a thermos of coffee, a miniature of whisky and a cheese sandwich for three hours with heavy snow outside waiting for what I had been told was going to be a very big litter. My reward for this devotion to duty – one very small piglet! I blamed the boar.

Preparation

Make sure that your sow or gilt has been well prepared for farrowing. In addition to separating her from other pigs some two weeks before farrowing you need to:

1 *thoroughly clean and disinfect her farrowing area;*
2 *treat against worms and parasites about seven days before farrowing;*

3 set a hurdle across one corner of your farrowing pen to form a secure area for your piglets. If you are using an infra-red heat lamp then set it above this area. Your piglets can get to and from the area under the hurdle but their mother can't; and ...

4 wash and clean her udders just prior to farrowing with warm soapy water and a little household disinfectant. This will ensure that she is free of any worm eggs that could have been laid around the teats of the sow and these eggs could then be transferred to the young piglets, causing many problems and resulting in the loss of piglets.

Knowing when

If you have followed the 'brimming' process carefully and noted when your sow or gilt was served then you can be pretty certain that farrowing will be around 112–15 days later. Separate your female at between 95–100 days and give her time to settle.

Watch for the following signs:

1 *Restlessness. Your female may become restless and may even try to break out in order to farrow elsewhere on or around the farrowing date.*

2 *Nesting. Pigs have a built-in desire to build a nest when farrowing. All pigs kept outside in a natural environment will gather leaves and branches to build a nest (similar to a swan's nest). This usually takes place on the day of farrowing or occasionally the day before. If your pig is not running outside in a natural setting then put some branches, dry straw or similar nest-making material in the corner of the paddock away from her house. If you see her carrying straw and branches to her ark then she is nest building and farrowing is close. She may, of course, build a nest outside which you will see easily. This is the time to encourage her to move indoors to farrow under cover and you should shut her in to make sure that she does not repeat the process.*

3   *Increase in size and shape of stomach and mammary glands.*
    Your sow or gilt will start 'bagging up' in the two weeks
    before farrowing and you should be able to see this change in
    shape quite clearly. This is another good indicator and time to
    separate if you have not already done so.
4   *Milk production. Shortly before farrowing your sow or gilt*
    will 'drop' or release milk and by gently squeezing one or two
    teats (rubbing her stomach at the same time helps) you will
    be able to get her to express milk. This can mean farrowing is
    very close (possibly that day). You should prepare yourself for
    the imminent farrowing but we have also known sows in very
    good condition produce first signs of milk up to three days
    prior to farrowing. Whatever the case may be, you do know
    that the event is close.
5   *The vulva becomes larger and also may noticeably redden,*
    while the muscles on either side of the tail slacken. This is a
    clear sign that farrowing is not far away.
6   *She will start contractions shortly before farrowing and*
    will breathe quite heavily (puffing and blowing) as she strains
    and pushes. Often she will push her back legs against the side
    of the wall or partition in her farrowing room in order to
    assist the pushing process as she lies on one side with her
    tail raised.

A sow lies flat on her side to farrow, pushing and raising her tail
as each piglet is expelled. Piglets can be born either head or feet
first and any that are reluctant to breathe should be held gently
upside down to drain mucus and fluids from the mouth and nose.
If a piglet does not breathe, gently but firmly rub the chest area
around the inside of the front legs and give as much warmth as
possible by placing under a heat lamp to dry out, or rub gently
but vigorously with a dry towel. Pushing a small piece of straw
up the piglet's nose if it is not breathing may cause it to cough
and start breathing. When you save your first piglet I can assure
you that tears will almost certainly roll down your cheeks. Warmth
is a very important factor when attempting to revive or save a
newly born piglet and a warm oven has been the saviour of many
a piglet.

If your sow or gilt is straining continuously and raising her legs to push yet nothing is happening don't panic, but if it continues for a long time then something is wrong. Call your vet.

Some breeders put the newly born piglets in a box full of straw until the whole process is over to protect the piglets from the sow lying on them while she is giving birth and exhausted. Generally I do not find this necessary as I feel the quicker the piglets suckle and get the mother's colostrum (first milk) then the better, although stronger piglets can be separated in this way for a short while if you have a number of weaker piglets that are being bullied away from the teats. Although not totally necessary, it is really good for you to be present as she gives birth. It is a memorable experience that you will never forget, plus the fact that your presence will enable you to assist a weak or vulnerable piglet.

The whole process may take several hours, so be prepared with warm clothes (especially on cold winter nights!) and a flask of hot beverage. Do not interfere with the farrowing process unless you have to or there appears to be a problem – the sow will usually cope quite well by herself. Simply make sure each piglet is breathing and that they are clear of mucus around the nose and mouth. Make sure the piglets are able to make their way to their mother's teats to suckle and get that all important first colostrum. Do not cut or interfere with the umbilical cord, although this is a good time to treat the cord with an antibiotic spray or dip.

Do not be fooled into thinking that it is all over if the pig stops producing after the first five or six piglets. She may just be resting and there may be more to come! You can only be certain that it is all over when she releases the final afterbirth. Beware because the afterbirth often comes along with a piglet that may suffocate if you are not vigilant. Once you are happy that it is over and all the piglets have suckled at least once, leave her to rest in peace after her exhausting experience. It is fascinating to watch the way she behaves with her piglets as she lies there resting but still allowing them to suckle. She will talk to her piglets who will immediately know that she is calling them to suckle and respond accordingly.

In all litters there are some piglets that are larger and stronger than others and it is therefore a good idea at this early stage to make sure that the smaller, more vulnerable piglets get enough colostrum. If you are present and you do see one or two weaker piglets struggling against the odds with their larger brothers and sisters then it is a good idea to remove the strongest for an hour or two. Provided that they have had an hour or more suckling you can do this without it affecting them, as long as they are kept warm and dry. The corner of the farrowing pen you have set aside for this very purpose with dry straw and a heat lamp is ideal provided you have a temporary barrier, such as a bale of straw, in front of the hurdle. This bale will stop the piglets whose suckling you are restricting from returning to their mother. During their absence you can physically hold the weaker piglets to the vacant teats to ensure that they get the sort of start to life that they deserve. It is a sad fact that more piglets die of starvation through an inability to suckle and get sufficient colostrum than for any other single reason. The smaller, weaker piglets are most at risk because they have to cope with having received a lesser transfer of antibodies from their mother via the placenta prior to being born and then have the problem of needing more colostrum to provide immunity than their stronger siblings.

Insight: Lining them up

I could never understand why a sow's teats had to be evenly matched until I saw my first farrowing. The first piglets born got the bottom row (mum lying on her side) but the last ones born had to stretch to the top row by sitting on the back of the one underneath, drinking from the matching teat. I could clearly see that this only worked if the top and bottom rows matched each other.

Once you are happy that the weaker piglets in the litter have suckled well you can remove the restricting bale of straw and allow all the piglets to suckle together. If necessary you can easily repeat the process. Keeping all your piglets warm is very important because you want them to build up their strength as quickly as possible by converting their food into growth and energy rather than just using it to keep warm. Make sure that your whole farrowing area is free from draughts and the use of an infra-red lamp in an area exclusive to the piglets will enhance your success rate as far as piglet survival is concerned. Not only will the infra-red lamp offer heat to your piglets but it will also encourage them away from their mother when they are not suckling, thus improving their chances of not being laid on. Piglets need warmth because the colder they are then the more energy they use up keeping warm and as they are born with limited supplies of energy they could soon become weak and die.

The importance of iron

If your piglets are born inside and do not have direct access to the soil you will need to make sure they get sufficient iron from an alternative source. When your piglets are born they will only have a limited supply of iron that will probably only last them for their first week of life at best. There will not be enough iron in your sow's milk to provide for your piglets' needs and unless they get additional iron from elsewhere they will die. There are several ways that you can deal with this problem:

▶ *An iron injection into the bottom is the surest way of being certain that your piglets have sufficient iron. Specialist iron mixtures for this very purpose can be obtained through your agricultural merchants but make sure they have some in stock well before farrowing takes place as they may have to order it specially. If you are injecting iron you should do so before your piglets are five days old.*
▶ *A clean turf of soil put in with your piglets every day in their secure area where the mother cannot get it will suffice, especially if sprinkled with iron sulphate.*
▶ *An iron mixture specifically prepared to be administered orally by placing at the back of the throat is an option but this should always be done within one day of birth.*

- ▶ *Iron licks can be placed in your piglets' secure area, as can iron in your piglets' drinking water but you may prefer administering the iron directly rather than leaving your piglets to administer themselves.*
- ▶ *Brushing iron sulphate paste on to your sow's teats every other day will further aid the intake of iron.*

If your piglets pass a white scour or dung then you can be sure that you have an iron deficiency problem so deal with it accordingly. If your piglets do not get enough iron they will be more likely to be sickly pigs that do not grow and develop as expected. Once your piglets have reached the age where they are happily eating creep feed they will be getting all the iron they need.

Castrating male piglets

The days when castration of all male piglets during their first week of life was a matter of course are now long gone. In my experience, it is totally unnecessary to castrate male piglets if they are being reared for pork and are taken to the abattoir before they are 32 weeks of age. Boar taint (a pungent and very distinctive smell that accompanies pork from an older entire male when it is cooked) usually only becomes a problem if your pig has been allowed to mate or is allowed to mature fully. Male pigs should not be used for producing bacon whether castrated or not and therefore the only time you might need to castrate your male piglets is when they are being sold for pets and not to breed. Although it is not that long ago when castrating a piglet was performed by the pig keeper it is no longer the case. If you need a piglet castrating it is very much a job for your local veterinary surgeon and not for you. After castration has taken place you must take every necessary precaution possible to ensure that there is no risk of infection. Put your castrated piglet in a clean, disinfected, dust free pen and do not let it out on pasture or in areas of dirt and mud until the wound has healed. A daily treatment with antibiotic spray or powder for the first three or four days should be sufficient.

Cutting teeth

As you may soon find out, your piglets' teeth are like little needles and are extremely sharp from the very moment of birth. This can, in some cases, cause trouble for the mother as they may damage her teats in such a way that infection sets in. It may result in mastitis and all the added problems that can bring. Despite the potential for problems there is no need to clip your piglets' teeth as a matter of course. In the vast majority of cases it is unnecessary and only causes the piglets stress they could do without. If you do not see any major problems in the first two or three days then it is best to leave well alone. I personally try to avoid cutting teeth. As your piglets grow and develop, so will their teeth and if you notice your sow not showing her usual enthusiasm for feeding her young she may be telling you something and it is time to wean the piglets. Provided that they are eating their creep feed there is no reason why you should not wean them from five to six weeks onwards in such circumstances and give mother a break, although eight weeks is my preferred age for weaning. If you do decide to clip your piglets' teeth then you should only cut the tips and under no circumstances cut down to the gums unless you want to encourage gum infection. Some good nail clippers will do the job but make sure you leave no jagged or sharp edges.

Creep feeding piglets

When first born and up to about two weeks of age your piglets will only be able to digest non-solid foods and cannot handle any weaner food high in fibre and starch. All creep feeds for very young piglets are primarily milk based because it is not until they are approaching three weeks of age that they will be able to satisfactorily digest non-milk proteins. This coincides almost exactly with the fact that your sow reaches her maximum milk yield three weeks after farrowing and then starts to decline. Because your piglets' need for nutrients is increasing at the same time your sow's milk yield is declining you will need to provide additional feed. Not only will you be producing stronger, healthier piglets but you will also be preparing their whole digestive system for weaning. Not every teat on your sow produces an equal amount of milk and therefore creep feeding your piglets from two to three weeks of age will compensate those piglets that are getting less milk from their mother. You can start putting a little specialized creep feed in your piglets secure area as early as four or five days after they are born and by sprinkling just a little over a clean turf or soil you might encourage them to start eating. By the time they are a week to two weeks old your piglets will be showing much greater interest in their creep feed and you can use the top of an old milk churn or the hollow top of a brick or heavy terracotta dish as a feeder that they cannot turn over. Do not leave any on the floor and feeding little and often is better than leaving your piglets with stale food – you should always pick up anything that is left over. If the feed you put down is fresh it will be much more acceptable to your piglets and will encourage them to eat. Always make sure there is an ample supply of fresh water.

In summary:

- ▶ *Fresh, clean water should be supplied.*
- ▶ *Empty out and throw away stale food.*
- ▶ *Encourage your piglets by feeding fresh, palatable food.*
- ▶ *Don't overfeed.*
- ▶ *Put feed where mother cannot get it.*
- ▶ *Increase creep feeding from three weeks of age.*
- ▶ *Give piglets on the teats with less milk a better chance.*
- ▶ *Little and often.*
- ▶ *Enable your piglets to wean more quickly.*
- ▶ *Terracotta trays or bricks make good creep feed containers.*
- ▶ *Stronger piglets will result as your sow's milk declines.*

Hand rearing

No matter how much you try, you may from time to time be faced with bottle feeding a piglet almost from birth.

The worst farrowing we ever had was when the sow had her piglets and then had a prolapsed womb (very unusual) which meant she could not suckle her piglets. I find bottle feeding one piglet fun, but ten every two hours is a different kettle of fish. By the time we had finished the tenth it was time to start all over again.

Before eventually taking the piglet away from its mother you should make sure that it suckles for as long as possible by being held to the teats to guarantee that it gets sufficient colostrum. Your piglet's chances of survival are greatly increased if it gains the necessary immunity provided by the colostrum. You can get sow's milk powder from your local agricultural merchants but you may have to buy a large sack when in reality you need only enough for one piglet. As mentioned previously, an excellent alternative is to use the best quality baby milk powder that can be mixed exactly as for babies. This will do the job just as well and will save you a great deal of money at the same time. Use a small kitten bottle or similar and feed about once every four hours, although in the first 24–48 hours they will do better if fed every two to three hours. Always feed at blood temperature and start with about 10–20 ml per feed and then gradually build up to 75–100 ml by which time you will only need to be feeding about three to four times a day. The sooner that you can get your piglet to drink directly from a bowl then the better it will be for both you and the piglet but it is even more important that you get your piglet on milk-based solid food as quickly as possible. You should try to do this when your piglet is about 7–8 lbs (3–4 kg) in weight because the longer you milk feed your piglet then the greater the chances become of it scouring. Getting your piglet started on a good specialized milk-based creep feed will help its digestive system mature at a faster rate and will also help you avoid overfeeding. It is very easy to overfeed piglets that you are bottle feeding or that are drinking milk supplement from a bowl. You must make sure that once your piglet is eating dry food there is plenty of clean, fresh water available for it 24 hours a day. When your piglet is a month old you can gradually start to wean it off its specialized milk-based diet and move it over to standard weaner food but, as is always the rule when changing diets, do it slowly!

Insight: Who needs congratulating?

It was two weeks before I realized that all the smiles and words of congratulation that I was getting whenever I went into town was misplaced. Someone had seen me in Boots buying Milton, a baby bottle, teats and baby milk and put two and two together making six. They didn't know I had a newly born sick and weak piglet that I was having to feed!

Some words of warning

Firstly, beware of making any piglets squeal because a squealing piglet could cause the mother to come to the defence of her newly born piglet. This could result in the mother treading on some of the other piglets and she could even attack you (even if she is normally very placid) if she thinks you are responsible for harming any of her piglets. Always have some means of protecting yourself in with you and always have a quick escape route. Take care and be alive to the fact that she will be very protective of her offspring.

Secondly, watch out for signs that your sow is finding it difficult to cope immediately and that she may be a 'bad mother'. This happens very rarely but occasionally a pig (more often than not a gilt) will attack its young, sometimes even eating them. If she seems to resent them while she is farrowing put them in a box or under the heat lamp until she has finished farrowing. If, when she has finished and you try to re-introduce them, she still shows signs of aggression then contact your vet for something to calm her down. This happens very rarely and in all my years of breeding I have not come across it personally but do know breeders who have.

Thirdly, make sure that any brick or concrete floor of your farrowing pen is well covered. You must achieve a balance between too much and too little. The piglets when first born are wet and can quickly die of hypothermia if they lie on a cold, hard floor for very long (concrete is the worst). Once they are dry and have suckled for a day or so then it is not so bad but that first 24 hours is crucial. On the other hand, if the straw or bedding is too deep or thick the newly born piglets will not be able to walk properly and may get stuck, resulting in them being trodden on because they cannot move out of the way. Although it is normal practice to leave an area of the floor free of straw (the dung and urine area) you must make sure that the entire floor is covered at farrowing time. Warmth is very important to young piglets and if they get cold as a result of insufficient, inappropriate or inadequate floor covering then they will die – beware!

Your new mother will want to drink plenty of fresh water (to help milk production) and her feed should be increased by around an extra 1 lb a day for each of her piglets up to a maximum of about 12 lb. This should be done gradually over the first few days and if she is clearing her feed well within 20 minutes, gradually add a little more. Beware of using water troughs that the little piglets could fall into, especially if the mother is moving her bedding around as it could pile up and act as a ramp for her piglets to climb up. Piglets are very inquisitive and active from a very early age and if they fall in a water trough they are sure to drown.

Finally, if you feel that your pig has been straining or in labour for a very long time then don't try to solve the problem yourself. Call your vet. It is better to take professional advice earlier rather than when it is too late. The same advice applies after the birth is completed. Watch your sow or gilt (as well as your piglets) for signs of ill health. Let the new mother get plenty of exercise and watch the way she behaves. If she does not eat normally, is very lethargic and she is constipated (there are no signs of dung in her pen) something is wrong and she could have 'farrowing fever' – call your vet. If you do not do so quickly, her milk will dry up and the piglets will suffer or, worse still, you could lose your sow or gilt.

As your piglets grow up, are eventually weaned and mother is taken away you will part with those that have been ordered as weaners and start to fatten those that are to eventually feed you, your family and friends. Don't think about running gilts and boars together for the rest of their lives if they are from the same litter. Separate them at around five or six months of age at the latest. You don't want your gilts 'in-pig' by their brother, do you? A boar can be ready for work from as young as five months and it is often only size that prevents him, while some young gilts will start 'brimming' around five or six months. The best advice we can give is to separate them and don't take the chance.

Notification, identification and registration

Not all pigs need to be registered but if you are breeding pure from a registered female and registered male then you should definitely 'birth notify' the litter with the BPA.

Insight: That's progress for you

When I first started there was no internet but now you can birth notify and register your pigs through the BPA website. You must notify all births within the first ten weeks.

A number of small pig keepers who are not serious breeders and are not considering showing their pigs often decide not to buy registered stock so registering all the offspring is an unnecessary expense for you. Your customers who buy pigs from you and don't want them registered get just as much pleasure keeping pigs by using them in ways that makes registration unnecessary:

1 *To mate with a boar of a different breed to produce 'cross-bred' piglets that can be sold locally to friends and neighbours who will fatten them for pork. They may even produce pigs to sell at the local market. They will benefit from all the good qualities in the traditional breed they started with.*

2 To simply fatten to produce pork for themselves, family, friends and neighbours. Many people do this as a 'starter' when keeping pigs for the first time and it may be the way you started. It does enable every new pig keeper to learn about keeping pigs, about the breed, whether they like keeping pigs and whether or not they want to breed registered pigs seriously in the future. Some new pig keepers try their hand at keeping more than one breed before making their final decision to breed registered stock.

3 As a long-term farmyard 'pet'. By 'pet' we do not mean it in the same sense as a cat, dog or pet hamster. A rare, minority or traditional breed of pig is an outdoor animal and should not be kept in the house. Many of our customers use their 'pet' pigs to rotovate the ground, clear a rough patch of land, keep woodland scrub down, as a means of disposing of excess milk from goats or cows or just because they like pigs on their land and only want to breed as and when they want to.

If you intend to enter the world of 'showing', if you intend to breed pigs seriously or if you intend to sell registered pigs, then registration is essential:

1 You must make sure that the parents, both the boar (sire) and the sow or gilt (dam) are both registered themselves.

2 If you buy adult stock make sure you get an official British Pig Association 'Certificate of Herd Book Registration' (Berkshire, Large Black, Middle White, Gloucestershire Old Spots, Tamworth and British Saddleback, Oxford Sandy & Black) or a 'Pedigree Certificate' (British Lop Pig Society).

3 If you are hiring a boar make sure that you see the appropriate certificate and ask for a copy if you are not offered one.

4 You must join the British Pig Association (BPA) or the appropriate breed society in the case of the 'British Lop' or 'Kune-Kune'.

5 Any litters you intend to notify must be ear marked as required by the individual breed society. Remember all pigs

must be numbered in sequence and it is customary to number males first, whether used for breeding or not.

6 *You must notify the BPA or the breed society (whichever is applicable in your case) of the birth of every litter that has been ear marked within ten weeks of birth. You should use the appropriate forms supplied.*

7 *When you have selected those pigs you wish to register (making sure that they meet the breed standards for registration) you must complete the Herd Book Registration Form and return it to the appropriate body. Remember that any stock leaving your premises unregistered may not be registered at a later date without your authority and signature so make sure that all your customers are aware of this.*

You will find the British Pig Association very helpful and we are sure that any breed society or breeders' club you join will give you all the advice you need.

Each new litter will always amaze and excite you because no two litters and no two piglets are the same. Each will have their own individual markings and personality. You may produce the perfect show-winning pig or just a really lovable character whose temperament and personality make up for any shortfall in conformation or marking. Whatever happens, your life will never be the same again.

Gestation calendar

If you know when your sow or gilt was served (S) then you can see at a glance when she is due to farrow. Do not rely solely on the size and shape of your pig to estimate when she is about to farrow because although many pigs show distinct signs of pregnancy early on in the gestation period, others fail to show many of the expected signs until very late in the pregnancy.

S	Due	S	Due	S	Due	S	Due	S	Due	S	Due
Jan	Apr	Feb	May	Mar	Jun	Apr	Jul	May	Aug	Jun	Sep
1	26	1	27	1	24	1	25	1	24	1	24
2	27	2	28	2	25	2	26	2	25	2	25
3	28	3	29	3	26	3	27	3	26	3	26
4	29	4	30	4	27	4	28	4	27	4	27
5	30	5	31	5	28	5	29	5	28	5	28
6	May 1	6	Jun 1	6	29	6	30	6	29	6	29
7	2	7	2	7	30	7	31	7	30	7	30
8	3	8	3	8	July 1	8	Aug 1	8	31	8	Oct 1
9	4	9	4	9	2	9	2	9	Sep 1	9	2
10	5	10	5	10	3	10	3	10	2	10	3
11	6	11	6	11	4	11	4	11	3	11	4
12	7	12	7	12	5	12	5	12	4	12	5
13	8	13	8	13	6	13	6	13	5	13	6
14	9	14	9	14	7	14	7	14	6	14	7
15	10	15	10	15	8	15	8	15	7	15	8
16	11	16	11	16	9	16	9	16	8	16	9
17	12	17	12	17	10	17	10	17	9	17	10
18	13	18	13	18	11	18	11	18	10	18	11
19	14	19	14	19	12	19	12	19	11	19	12
20	15	20	15	20	13	20	13	20	12	20	13
21	16	21	16	21	14	21	14	21	13	21	14
22	17	22	17	22	15	22	15	22	14	22	15
23	18	23	18	23	16	23	16	23	15	23	16
24	19	24	19	24	17	24	17	24	16	24	17
25	20	25	20	25	18	25	18	25	17	25	18
26	21	26	21	26	19	26	19	26	18	26	19
27	22	27	22	27	20	27	20	27	19	27	20
28	23	28	23	28	21	28	21	28	20	28	21
29	24			29	22	29	22	29	21	29	22
30	25			30	23	30	23	30	22	30	23
31	26			31	24	31	23				

(Contd)

S	Due	S	Due	S	Due	S	Due	S	Due	S	Due
Jul	Oct	Aug	Nov	Sep	Dec	Oct	Jan	Nov	Feb	Dec	Mar
1	24	1	24	1	25	1	24	1	24	1	26
2	25	2	25	2	26	2	25	2	25	2	27
3	26	3	26	3	27	3	26	3	26	3	28
4	27	4	37	4	28	4	27	4	27	4	29
5	28	5	28	5	29	5	28	5	28	5	30
6	29	6	29	6	30	6	29	6	May 1	6	31
7	30	7	30	7	31	7	30	7	2	7	Apr 1
8	31	8	Dec 1	8	Jan 1	8	31	8	3	8	2
9	Nov 1	9	2	9	2	9	Feb 1	9	4	9	3
10	2	10	3	10	3	10	2	10	5	10	4
11	3	11	4	11	4	11	3	11	6	11	5
12	4	12	5	12	5	12	4	12	7	12	6
13	5	13	6	13	6	13	5	13	8	13	7
14	6	14	7	14	7	14	6	14	9	14	8
15	7	15	8	15	8	15	7	15	10	15	9
16	8	16	9	16	9	16	8	16	11	16	10
17	9	17	10	17	10	17	9	17	12	17	11
18	10	18	11	18	11	18	10	18	13	18	12
19	11	19	12	19	12	19	11	19	14	19	13
20	12	20	13	20	13	20	12	20	15	20	14
21	13	21	14	21	14	21	13	21	16	21	15
22	14	22	15	22	15	22	14	22	17	22	16
23	15	23	16	23	16	23	15	23	18	23	17
24	16	24	17	24	17	24	16	24	19	24	18
25	17	25	18	25	18	25	17	25	20	25	19
26	18	26	19	26	19	26	18	26	21	26	20
27	19	27	20	27	20	27	19	27	22	27	21
28	20	28	21	28	21	28	20	28	23	28	22
29	21	29	22	29	22	29	21	29	24	29	23
30	22	30	23	30	23	30	22	30	25	30	24
31	23	31	24			31	23			31	25

10 THINGS TO REMEMBER

1 *Three is the magic number. The gestation period is 3 months, 3 weeks, 3 days, 3 hours, 3 minutes, plus 3 glasses of wine = piglets!*

2 *Three is the magic number. A pig comes into season (brimming) and is ready to be served every 3 weeks.*

3 *Three is the magic number. A pig is in season (brimming) for 3 days.*

4 *Always worm your pig 10–14 days before farrowing to stop placenta crossover of worms.*

5 *Increase the amount of pig nuts you give your sow by 1 lb a day for each piglet born. Spread the increase over the first seven days.*

6 *Spray the umbilical chords with Terramycin aerosol spray to stop infection.*

7 *Make sure that the mother eats and drinks well within a maximum of 24 hours after the birth of the last piglet.*

8 *Test for milk by 'milking' the mother twice a day for the final seven days. When you get milk she is going to farrow within 24 hours but if there is a lot of milk maybe sooner. Pigs tend to farrow when it is quiet and often at night. Although it might sometimes seem like it, they do not always farrow at 2 a.m. in the morning! The moment she produces milk, clean her 'underline' (teats) with warm soapy water and a little disinfectant.*

9 *Keep a note of your pigs' 'brimming cycle' before using artificial insemination (AI).*

10 *Be present at the birth of your piglets and smile more than ever before.*

9

Meat for the freezer

In this chapter you will learn:
- *how to rear pigs for the table*
- *about providing meat for family and friends*
- *how to select a good abattoir and suitable butcher.*

How heavy is your pig?

Without knowing the approximate weight of your pig you will be unable to decide whether or not it is ready to go to the abattoir. How can you best tell?

Buying the correct pig scales can be expensive with new ones costing many hundreds of pounds. Even if you see some at a farm sale they may cost more than you can justifiably spend if you only have two or three pigs. You could, of course, take them to neighbours if they have the facilities but moving your pigs back and forth is probably another extra task that you could do without.

Don't despair because there is another, far cheaper way.

- ▶ *First measure your pig's length in inches from between the ears to the root of the tail.*
- ▶ *Next measure the girth in inches around the shoulders, tight to the pig's front legs.*

► *Multiply the length by the girth and divide by 12 if it is a lean pig, 11 if it is medium and 10 if it is fat. The answer is in pounds.*

Example: Length = 50 inches
            Girth = 40 inches

Length (50) × Girth (40) = 2,000
2,000 ÷ 10 (assuming the pig is fat) = 200 lb

To measure the pig use a piece of string or baler twine (the farmer's friend) and tie a knot at the appropriate point. So you can then measure this later to work out the weight. This method will give you a good guide but remember that it is only a guide not an exact weight but for your purposes it will more than suffice. It will tell you whether or not your pigs are near or approaching porker weight. If your pigs are around 160 lb then they are about ready for the abattoir as porkers.

As a rule of thumb 'carcass' weight is usually about 33 per cent less than the live weight. This is slightly more in younger, lighter pigs (say, less than 100 lb live weight) and slightly less in very heavy, older pigs (say, over 300 lb live weight). Naturally the weight loss is greater if a pig is fed shortly before slaughter and the general condition of the pig does have an effect. Do not feed your pig on the day you are going to the abattoir.

The ideal weight

What weight should your pig be and how long should it take? These are two questions that everyone considering keeping pigs for the first time wants answered if planning to rear pigs for the table. As with many things in the world of pigs there is no definitive answer because it does, of course, all depend on the conditions that your pig is being reared under. It also depends on what sort of final end weight you or your customers want.

If you aim for a 'dead weight' of around 120 lb (54.43 kg) then you will get some great meat but 100–40 lb (45.36–63.5 kg) is perfectly acceptable for a porker. Of this 'dead weight' allow around 20 lb (9.07 kg) for head, trotters, tail and the unusable pieces and this will give you a final usable (saleable) weight of about 100 lb (45.36 kg), making the total weight of half a butchered, wrapped, labelled and boxed pig about 50 lb (22.68 kg).

If you are fattening your pigs for your own freezer you can decide the weight you want but seek the advice of your butcher or abattoir if in doubt. Some butchers and abattoirs do not like porker pigs to be too heavy and you may in any case want your pig at a heavier weight for bacon of about 140–70 lb (63.5–77.12 kg) dead weight or 200 lb (90.72 kg) plus live weight.

A pig that is kept running outside (free ranger) will always fatten more slowly than a pig kept in intensive farming conditions. You can expect your pigs that run freely outside for most of their lives to reach porker weight of about 150 lb (68.4 kg) live weight by about 26–30 weeks of age, whereas commercial pork producers would aim for this weight in half that time or even less. You will not have the extra costs of lighting, heating, high protein feeding, commercial housing and all the other costs involved in an intensively reared operation and the quality of your pork will send shivers of delight down your spine.

Insight: In at the deep end

I am often asked if pigs can swim and the answer is a resounding 'yes' but I was taken aback once when someone on one of our courses said that they could even swim after they had been to the abattoir and then cured. This course member then explained that whenever he put commercially reared bacon in his pan the first thing it tried to do was 'swim out' in order to avoid drowning in the water that oozes out. There was a lot of nodding in agreement that day.

More importantly, in my opinion, is the fact that you have given your pig a free and better life. You have allowed the pig to develop naturally at its own pace, producing a good layer of fat that gives the true flavour and texture to the meat. It is now recognized that by breeding very lean carcasses much of the real flavour is lost and this results in pork that so often has no more flavour than cardboard. All too often the commercially reared pork that is available in supermarkets today relies for its taste on the apple sauce, gravy, creamy buttered potatoes, herbs, stuffing and condiments you might add to it.

Insight: On the 'wanted' list

In 1970, when I sold my first five halves of rare breed pork, I suddenly started acting like a criminal on top of the 'most wanted' list. Every time I went into town I skulked in and out of stores hoping I would not meet anyone who had bought my pork. I simply didn't really know what they would think of it. Only when I was pounced on in the dark recesses of a supermarket by someone screaming 'There he is – the amazing pork man' did I realize that everyone wanted this amazing taste in their freezer. I was now a free man.

Some people may find it difficult to accept that they are going to eat their own pigs. However, if you are considering fattening for your own freezer and have any doubts, remember:

▶ *You have given your pigs a better and much longer, happier, stress-free life than they would normally get.*
▶ *You are going to have meat that has been reared without being 'forced' and that does not contain growth hormones, chemical additives, etc. You know what your pig has eaten and how it has been reared.*
▶ *You are certainly helping to conserve rare and minority pig breeds, especially if you are fattening young hogs for meat.*

You may well end up with customers who buy their pork from you but who could not kill and eat their own stock – you should accept and respect their feelings. Some people may find it difficult to accept that eating rare breeds helps conservation. Initially this may appear to be a contradiction in terms, however, it is a fact. It is only by finding a market for meat from surplus stock (not all piglets from a litter can be used for breeding) that breeders can afford to continue keeping and breeding the traditional rare and minority breeds. While gilts may be sold for breeding if they meet the required breeding standards of a particular breed, there is no need for so many boars. One good boar will be able to sire many gilts and sows in his lifetime and therefore in most litters the boars are surplus to requirements as far as breeding is concerned.

It is not possible for you or any breeder to keep and feed excess stock as it will lead to financial ruin and therefore it is imperative that a market is found for quality premium rare breed meat. There are only two choices for the breeder if they do not find such a market – either stop breeding or stop keeping pigs. Either way it reduces the chances of us protecting our rare and minority traditional breeds.

Insight: Sausage mania

The first solid meat sausages my partner made were so popular that we sold out (enough to normally last three weeks) in less than two hours at a local show. Check for yourselves how much 'pork' meat is in the recently introduced 'Great British Banger' and you will be as shocked as I was.

Your breeding stock will be ready for the boar by nine months old and should have a weight of about 200 lb (90.72 kg). At this weight your pig should be good, fit, lean and not too fat. Too much fat will mean that you may have difficulties getting your pig to conceive. Your own fattening porker pigs should weigh around 150 lb (68.4 kg) at about six months of age if they are well fed in preparation for the freezer. Provided they are not overfed and they are well exercised they will not put down excess fat and you will get succulent, sweet and tender meat that tastes as pork really should. Feeding lots of apples in the two weeks prior to taking your pig to the abattoir does not mean you can do without apple sauce at a later date. Remember that a good layer of fat (but not excess) adds flavour to the meat throughout its growing life.

A pig for sale

If you are selling your pigs then your customers – whether butchers, restaurants or direct to your friends and neighbours – will determine the weight they want. If selling direct to friends and family always sell a fully butchered, wrapped, labelled and boxed half pig fresh and ready for the freezer. Specific joints will be dearer, a lot more trouble and you will probably be left with the joints nobody else wants. You should sell your meat direct from the butcher who prepares the meat and you must not touch, wrap or box the meat yourself unless you want to tackle all the hygiene rules and restrictions you will be required to adhere to. Let the professionals (butcher and abattoir), who are licensed to prepare meat, act on your behalf. Remember that a well-fed and correctly kept traditional old style, free range pig produces a succulent meat of high quality and unique flavour so don't ruin everything at the last minute by trying to prepare or package the meat yourself.

Don't forget that an outside pig, running free and not constrained in a sty, fattens at a far slower rate than one kept inside with restricted movement. Generally, the front end of the pig should be relatively light (head, neck, jowl and shoulders) as there is little sales value in these parts. The back should be long and the hams broad and firm. Long thin, narrow bodied pigs that are 'leggy' are not good pork pigs, although they may make great bacon.

If you intend getting a sensible return from your pigs then putting 'all your pigs in one basket' is not a good idea. It is better to be able to offer weaners (good pure bred piglets) as a prime source of pig income, 'farm to freezer' porkers (direct to friends, neighbours and general public) next. Whatever you do with pigs there is no fortune to be made as a small-scale pig keeper no matter what anyone tells you. However, you can more than cover costs and show a fair return if the system is managed correctly. More importantly you will have wonderful meat in your freezer and great pleasure from your piggy experiences.

Insight: The 'real thing'

I was most hurt when a relative told me our sausages were rubbish until they phoned back in the evening to confess that, unbeknown to them, their teenage son had substituted supermarket sausages he should have taken camping so he could impress his girlfriend with what he called 'the real thing'!

Sending your pig to the abattoir

As hard as it is, sometimes you will always be faced with sending your pig to the abattoir (whether you take them yourself or arrange for someone else to assist). If you ensure that your pigs have a very good quality of life while they are with you, far better than they would have under normal conditions, then the journey to the abattoir will be a little easier. Try to ensure that when the end comes for your porkers and baconers they are treated as kindly as possible.

You should make certain that you give your pigs a good quality of life no matter what their final fate might be. When it comes to 'freezer time' choose your butcher and/or abattoir wisely. I believe that the killing of any animal for food should be done as kindly and humanely as possible. A quiet environment with an individual pen is ideal and so much better than having your pigs driven along (conveyor belt fashion) with hundreds of other pigs who have had a less fortunate life than yours.

Selecting a smaller, more personal, butcher or abattoir will also help ensure that you actually get back the same pig that you took in for killing. I have heard numerous stories of the wrong pig being returned from larger, less personal, abattoirs and although I have had no personal experience of this there is very rarely smoke without fire. Whether this is because of carelessness, because the abattoirs think 'a pig is a pig is a pig' or because they deliberately wanted the free range pork for themselves I do not know, so choose your abattoir and butcher wisely. Killing, butchery, wrapping and boxing ready for the freezer is far less expensive than you might think.

If a pig (or any animal for that matter) is badly stressed at the time they are dispatched there is an increased flow of adrenaline which is liable to cause a change in the way the meat finally tastes. We believe that although it is necessary to breed and sell meat in order to save the rare breeds it is equally right to see that they are dispatched as kindly and humanely as possible. Try not to leave your pigs at an abattoir overnight in strange surroundings and among many other strange animals, especially if there is the 'smell of death' around. It is not necessary as many small abattoirs are more than pleased to dispatch the same day, often within minutes of you delivering your stock. Try also to find a small private abattoir as the larger, more commercial units with thousands of animals a day passing through simply do not have the time to 'care' as you would want and your pigs are sure to be stressed. You also need to be certain that you get your own 'premium, high-quality, naturally fed and reared' pork back and not some commercial carcass by mistake.

Insight: Too much noise

Someone once told me that you couldn't eat naturally fed and reared pork while listening to the radio at the same time. Why? 'Too much crackling!'

Make sure that when you take your pigs to the abattoir they are not put in a pen with other pigs for they will inevitably fight each other and try to establish themselves at the top of the pecking order. This fighting is bad for your pig and bad for the meat because it can cause capillary bleeding, which will affect the look of the meat.

Load your pigs into the trailer the night before they are due to go to the abattoir and you will save yourself a lot of heartache and extra trouble. They will load more easily at feed time the night before and will happily spend the night in the trailer as well as being less stressed on the day they go.

Insight: More haste and less speed

I used to find it infuriating when I was in a hurry to get to the abattoir for an early morning place in the queue and my otherwise easy-to-deal-with pigs played me up and were difficult to load. It turned out to be because I was changing their routine by getting them up early and not feeding them. I now load them into the trailer the night before at their normal feed time and let them sleep in the trailer overnight. Much easier and another blissful hour in bed the next day.

Do not feed your pigs in the morning but do give them water. Organize a good butcher (if he also has the abattoir so much the better) to cut and prepare your meat for you. Make it clear how you want it done and get your butcher to wrap, box and if possible label the cuts for you. Presentation is all important and you do not want to find yourself collecting a black bin liner full of bits of unwrapped meat from all your pigs.

Remember some pig breeds make better pork and some breeds make better bacon:

- *Dual purpose pigs that are suitable for both pork and bacon include the British Lop, British Saddleback, Tamworth, Oxford Sandy & Black and Gloucestershire Old Spots.*
- *When it comes to specialist pork pigs the Middle White and the Berkshire reign supreme. Although the Kune-Kune is primarily a 'pet' it is eaten in its home country of New Zealand as it is throughout Europe.*
- *The Large Black produces a long, lean bacon carcass and although it produces first-class pork it is more of a bacon pig and is outstanding when grown on to be cured.*

Insight: Where does that come from?

The expressions 'pig in a poke' and 'the cat's out of the bag' come from the days when piglets were sold by putting them in a sack ('poke') and hanging the sack on a market stall. People would then feel the sack, count the piglets and buy 'a pig in a poke'. Unfortunately some rogue traders slipped in a kitten or two to make the sack look bigger and it wasn't until they got them home that the buyers realized they had been duped because 'the cat was out of the bag'!

10 THINGS TO REMEMBER

1 Weigh your pig with a piece of string!

2 Get a licensed abattoir to kill your pig and a licensed butcher to cut, wrap, label and pack the joints for you.

3 When selling your meat do not freeze it. Arrange for your customers to collect it on the day you get it back.

4 A half pig that is jointed, wrapped and boxed does not take up much freezer space.

5 Prepare your customers in advance by telephoning them a week before it is ready.

6 Use a small abattoir that will dispatch quickly and won't mix your pigs with others so they fight.

7 Load pigs at feeding time the night before they are going to the abattoir.

8 Remember that the only way we can save our rare breed pigs is to eat them.

9 Do not feed your pig on the day it goes to the abattoir – water YES, feed NO.

10 Never forget that producing quality pork and bacon depends on your pigs being naturally fed and naturally reared, as well as on good husbandry.

10

Sales and marketing

In this chapter you will learn:
- *how best to sell and market your pigs*
- *about advertising your pigs*
- *how to get that all-important customer visit.*

Having bred your first piglets the delight I know you will feel at seeing a good healthy litter is one thing but selling your surplus stock at a reasonable and cost-effective price is another. If you are to become a successful, long-term pig breeder then you must not allow your emotions to override common sense. As wonderful as they are, you simply cannot keep every piglet you ever breed or they will eat you out of house and home.

If you are to become a successful small-scale pig keeper then not only must you develop good husbandry skills but you must also learn some of the basic business and sales skills that will ensure the success of your small pig business. Whether it is just a part-time hobby that you do simply for pleasure, or if it is something more, you must still make sure that you do not become overrun with pigs. If this happens then not only do you and your family suffer but so do your pigs.

Insight: The most important sale of all

I used to think that the first sale was the most important, as was every new customer. It wasn't long before I understood this was not the case because if they never came back I was chasing my own tail. The most important sale is the second one because 'repeat orders' are what you build a successful business on.

Many people believe that it is just the art of breeding that will save our rare and minority breed pigs but this is only part of the story because there is much more to it than that. Breeding without being able to establish a market for them (whether for further breeding, for 'pets' or for meat) is even more disastrous than not breeding at all. If you are forced to kill your adult breeding stock because you can no longer maintain the herd then another pig in that breed line is lost forever.

If you are in any doubt about your ability to find a market for your piglets you might like to breed then stand back and think again. Think very carefully and remember the following:

▶ *It is not the breeding of the rare breed piglets that will save our rare breed pigs.*
▶ *It is not the fattening of the rare breed porkers that will save our rare breed pigs.*
▶ *It is not the caring for and loving of our rare breed pigs that will save them.*

While the breeding, fattening and caring for our rare breed pigs might help and will certainly play an important part in any small-scale pig keeper's life, it is the ability to be able to market and sell them (at a cost-effective price) that will truly ensure their long-term future. Our rare breed pigs became endangered because the breeders were not able to sell their stock either as weaners, growers, adult breeding stock, porkers or as meat at realistic cost-effective prices. They stopped keeping them because they could not afford to subsidize an animal they loved. Sad but true.

Insight: Think longer term

Buying cheap is as bad as selling cheap. If you buy piglets
for less than you know they cost to produce you are harming
yourself as well as others. Firstly, you are contributing
towards a loss-making operation, ultimately sending the
breeder out of business – then where do you go for your next
piglets? You are also probably contributing to the demise of
the rare breed pigs because if that breeder cannot afford to
keep going, because the price you are paying is not sufficient
to make breeding financially viable, then there is one less
to promote the breed. Secondly, you are helping establish
an unrealistic selling price for yourself. Thirdly, if you have
to pay more next time then your selling price has to go
up to compensate for this, thus upsetting (or losing!) your
established customers.

Base philosophy

It is easy to give piglets away – any fool can do it. There is always
a temptation to 'sell at any price' in order to make the sale but you
must understand that this is counterproductive. It does nothing for
the rare breeds (which in my opinion are undoubtedly superior in
all ways to normal commercial pigs). It also does nothing for you
the breeder except help lose you money and perhaps make you a
long-term member of the 'I loved pigs but couldn't afford to keep
them' club!

Insight: Any fool can give things away

I have never met a successful pig keeper who sells piglets for
less than they cost to produce but I have met a lot of failed
pig keepers who have. I cost everything to the last detail and
so must you.

Although many commercial farmers may complain about the
price of pigs and then bemoan the fact that 'there is no money in

pigs', there is more than one side to the story. What is their sales and marketing strategy really like? Most farmers have never sold anything in their lives – they simply take their stock to market and let someone else sell it for them and then pay the auctioneer a commission for selling at what is more often than not a non-cost-effective price. This simply doesn't make sense. What makes matters even worse is that the market charges they incur (transport, auctioneer's commission and ancillary charges) go up each year even if prices fall. It is called 'market forces' – factors outside your direct control that could affect either the end selling price or price offered to you. It is important that you keep control of your sales and marketing rather than hand the future of your pigs lock, stock and barrel over to someone else.

When you sell rare and minority breeds, or any pig that has been kept, fed and reared naturally then you are selling something very special whether the pigs are intended for breeding or for fattening. You should never think you are competing with commercially 'factory farmed' or intensively bred and reared white breeds because you are not. If your potential customers want the best they will buy the old traditional breeds and naturally kept pigs that have been free range reared. Anyone who wants the sort of quality that comes with naturally fed and reared pigs will be prepared to pay a fair and reasonable price for it.

Insight: Don't compete – make yourself better than elsewhere

We know that every butcher and supermarket sells pork, bacon, sausages, ham and gammon so we decided not to compete years and years ago. We sell 'A mouth-watering eating experience'; we sell 'Crackling that crackles like it used to'; we sell 'Joints that don't shrink in the oven'; we sell 'Bacon that doesn't swim out of the pan'; we sell 'A dream of the past'; we sell 'Juices that dribble down your chin just thinking about it'; and we sell 'Every eating sensation that the supermarkets and butchers don't'. In this way we are not in competition with them – you should do the same and avoid the competition.

You are wrong if you think that the 'stack it high and sell it cheap' approach works with pigs – it doesn't. All that happens is the breeder loses money plus the breed loses both status and real value in the eyes of any potential customers or new pig keepers. Price, as far as rare breeds and naturally reared pig buyers are concerned, is not the most important issue. The qualities your pigs possess are far more important than price.

Rare breeds and naturally reared pigs generally possess the following important qualities:

▶ *A great temperament that makes them easy to handle, which is of particular benefit to the new pig keeper.*
▶ *Superior meat quality when bred pure, fed and reared naturally. This was proven in a study carried out by the University of Bristol in which they carried out controlled blind taste tests to assess 'the quality of pork from traditional pig breeds'.*
▶ *Easy to keep because they are hardy and make ideal 'outdoor' pigs so they can be kept without the pig keeper having to go to the great expense of building specialized pig units.*
▶ *An attractive appearance. Colour and markings often play an important part in the minds of the new pig keeper when they are selecting their first pigs. Most small-scale pig keepers want more than just any old 'pink' pig – they want something different. This is probably something that you can easily relate to.*

Be sensible with your marketing and plan ahead. In doing so you will avoid selling your rare breed pigs at a loss. There is nothing to worry about when it comes to asking a fair and justifiable price for all your stock if in return you give your customers something special as well as something extra. When buying from you, your customers should not just get a first-class pig but they should also get all the help and advice you can give them. This will result, in many cases, in a long-term 'piggy friendship' because you will have so much in common. If you aim to make breeding pigs a fun and enjoyable pastime then you will find that you will make many friends as well as meet a lot of very nice people.

Selling your piglets

Don't think that selling cheap weaners encourages people to become long-term breeders and pig keepers because it doesn't. Asking a realistic and sensible price for your weaners helps ensure that the person making the enquiry gives thought to the purchase. It is more likely that they will take greater care of their new pig than someone who has bought on the spur of the moment because their investment was minimal. If they are the right people to keep pigs, they would still have bought at a more sensible price. It will be the breed and its qualities combined with actually having seen your delightful piglets that make people buy and not any 'give away' offer that you may come up with.

Once a weaner is seen in its litter situation then it sells itself (and often another one or two to go with it). The first aim must therefore be to convert as many enquiries to the 'visit and see' situation. If you are really enthusiastic about pigs, rare breeds and saving the endangered breeds from extinction then you should demonstrate this in your voice when answering queries. Enthusiasm is contagious.

Insight: Brush up on your spelling

Most people know that 'sell' is a four-letter word but you should try and spell it differently: I know the successful way to spell 'S-E-L-L' is 'H-E-L-P'. It is only by making sure you demonstrate that buying piglets, bacon, pork or ham from you is going to 'benefit' (help) your customers in one way or another that they will buy and then also become regular customers. If you just 'sell' for sales' sake and what you are selling is of no benefit to them, you will fail so spell 'S-E-L-L' the same way as I do.

You have no need to actually 'sell' your pigs in the manner of the double glazing or telephone salesperson because your piglets will sell themselves if you follow a few basic rules of marketing. The fact that your piglets are so soft, warm and silky should be a constant reminder to you that you are not in the world of the 'hard' sell.

RULE 1 – THE FOUR LETTER WORD

As said above, you need to understand that in your mind S-E-L-L it must be spelled H-E-L-P. If you are not able and willing to give all the help and advice you can to anyone thinking of owning a pig then you will become just another ordinary pig breeder who makes one-off sales. Anyone who is successful in business will tell you that real success in any business comes from two things:

1 *Repeat orders.*
2 *Word of mouth recommendations.*

Unless you are prepared to help everyone you deal with they are unlikely to return again and will certainly not spread the word that you are the person to go to if they want rare breed and traditionally reared pigs in your area. You should simply want people to visit you and see your pigs in the natural environment in which you keep them. If you believe that honest help, advice and friendship are what prospective pig keepers want then you can leave the 'selling' to the pigs themselves. Those delightful, soft, warm and cuddly piglets will do a better job than you will anyway.

Insight: Piglets are not piglets for long

Although we don't let our piglets go until they are properly weaned (eight weeks) and have spent a further week away from their mother (nine weeks) we make sure that we sell all our piglets before they are three weeks old. We advertise them for sale even before they are born so prospective customers see them when they are cute, adorable piglets. They are far less attractive when they are older and that 'Aaah!' factor diminishes, making them harder to sell. We let our piglets sell themselves.

RULE 2 – AIM FOR A VISIT, NOT A SALE

Do everything you can to ensure that anyone showing interest actually visits to see the piglets. What stops a person from visiting? Three things can play a part in stopping that all-important visit and they are the price, trying to sell on the telephone and your location.

Price

People are often put off if they are given a price in isolation without being able to visually relate to your piglets. In any litter of piglets there will be at least six different prices. Gilts are more expensive than boars, registered pigs are more expensive than unregistered stock, a potential show pig that meets perfectly all the breed standards and has a good underline will be a different price from a runt. It is therefore impossible to give a price over the telephone because if you give the price of the dearest pig in the litter it will probably seem too much but if you give the price of your cheapest piglet your potential customer could possibly select your most expensive piglet litter and expect to pay the quoted 'cheap' price. You must therefore be absolutely honest and explain why giving a price is impossible until they have seen the piglets. Once you get them to come and see the piglets they will sell themselves and price is irrelevant. If pressed then give as broad a range as possible, leaving all your options open. How can you give an exact price when you don't know which of the litter they will choose and so many other factors affect the end price?

Don't forget that giving a firm price on the phone could affect your sales in three ways:

1 *You could lose the person who thinks the piglets are underpriced.*
2 *You could lose the person who thinks they are overpriced.*
3 *You could lose the extra money from the person who would have paid more when they actually saw the piglets but you fixed the price too early.*

Insight: The answer to success/failure is staring right back at you

If you make a successful sale or get a new long-term customer and want to know why, just look in the mirror. Or, if you fail to convince someone your pork is the best and they go somewhere else, you must still look in the mirror. Both success and failure is down to the person you see in the mirror so don't blame your pigs, just work harder getting the things you do right!

Selling

It is often selling on the telephone that stops someone from visiting so you must curb your enthusiasm to get an order for your piglets. Don't try to sell on the phone, just be pleasant and help them. If you handle the enquiry properly your weaners will do the selling for you so why make life difficult for yourself?

Always make your aim to 'help' because helping is a form of giving. When you give something to someone it is human nature that they want to give you something back. Your aim should be to make the something they want to give you, in return for you being so helpful, a visit to see your piglets. After all, even if someone visits and doesn't buy you have had the pleasure of talking pigs (your pigs) and will probably have made new friends. The important point being that you will have enhanced your reputation and 'word-of-mouth' advertising is always the best.

Remember that you have all the aces when it comes to selling your rare breed piglets. The very largest majority of people who contact you will know nothing (at best very little) about your particular breed. You are the expert and therefore the best qualified to give help. Putting out a short list of dos and don'ts plus feeding instructions takes very little time but will be invaluable to your new customer. It will ensure your piglets are well cared for in their new home.

Do offer to send or email breed information and a photograph along with simple directions on how to find where you are located. Remember that many of your telephone callers may not even know what some of the breeds look like. Even if they know the breed they won't have seen anything as pretty as your piglet. Sending something 'free' is very beneficial because it portrays you as a 'helper' not a 'seller', which subconsciously breaks down barriers. They will instinctively feel that coming to see your weaners is something to look forward to and not something to fear.

Don't attempt to seek a sales commitment on the telephone. Instead use phrases such as:

- *no obligation*
- *bring the children for a fun day out*
- *just come and look – we love showing people our pigs. Whether you buy a piglet or not is immaterial. I will be delighted to give you all the help and advice I can.*

These phrases are all designed to get people to visit without feeling that there is any pressure on them to buy your piglets. You can be confident and know that as soon as they are looked at, your piglets will inevitably sell themselves.

Location and distance
If you make coming to visit you sound difficult, complicated or a long arduous journey then people will not want to come and see your piglets. Make your location sound as close as possible to the

caller and easy to get to. Here are some dos and don'ts that might help you when giving your location:

▶ *Don't talk mileage because 40 miles soon becomes an 80 mile round trip in the mind of your potential customer.*

▶ *Do talk time from the nearest major town (X is only just over 30 minutes from us).*

▶ *Don't give telephone directions – offer to send them instead. Telephone directions make things sound complicated and you will want their address anyway for your records. It may mean spending on postage or taking a little time writing an email but if you sell your weaner piglets at the right price then it is worth the extra effort. Send standard instructions on how to find you. It is more professional and receipt of a letter or email from you adds weight to their commitment to visit. Remember that a visitor to see your piglets almost always ends up becoming a buyer.*

▶ *Don't make your potential customer sound the exception when it comes to distance if they are coming a long way. If true, use phrases like 'We have people to see our pigs from all over the country' and if you have sold to anyone in their area or general direction (especially if it is even further away) then mention it. You must always make people like to feel they are not alone. Joining in the rush to have a good breed of pig is more desirable than feeling the odd one out.*

▶ *Remember the customer's needs come first, not yours! It will serve you better in the long run if you change your plans to accommodate someone's visit rather then asking them to change theirs to satisfy you.*

▶ *Even if people say they are coming tomorrow, ask for contact details and follow up with a telephone call or email prior to the visit. This second contact is important because it 'firms up' the visit and also shows you to be a caring, helpful person. In the same way as good restaurants ask for your telephone number when you make a booking, you should commit anyone proposing to come and see your pigs or piglets – there is nothing worse than giving up time to show someone your stock and they don't arrive!*

▶ *Do make the directions to come and see your piglets as simple and as easy to follow as possible.*

RESPONDING ON THE TELEPHONE TO THOSE
ALL-IMPORTANT ENQUIRIES

Have you ever stopped to think just how difficult it is for someone to pick up the telephone in answer to an advertisement? Phoning a complete stranger is not easy. The person at the other end of the telephone does not know what you are like, what the response will be and very often doesn't know what to say, apart from 'I'm phoning about your advert'. Your first job is to make the caller feel as much at home as possible. A good way to keep this firmly in your mind is to have a card, with the word FARMER printed on it, right next to the telephone:

Friendly
Amusing
Responsive
Merry
Enthusiastic
Relaxed

Make sure you behave in exactly that way when you respond to any enquiry.

Do not be too formal, abrupt, curt or overpowering. If you can make the person making the enquiry laugh then he or she will be more relaxed and you will be better able to build a relationship on the telephone that will result in a visit.

When you get a call, endeavour to find out if the potential customer has kept pigs before. You will obviously take a different approach with a novice pig keeper to the one you would take with someone who has kept pigs longer than you have.

Armed with this information you are in control and ready to start helping as you know a little about the needs and interests of your potential customer. Asking a question is also a good way of by-passing the issue of price as very often the question 'how much?' is either forgotten or comes lower down their list of priorities.

You can extol the virtues of your breed (and of keeping pigs generally) to your potential customer no matter what their experience of your breed. If they don't know much then you can tell them about the wonderful nature, wonderful appearance, ease of handling, cleanliness, intelligence, great tasting meat, ability to be wonderful mothers, etc. If they do know the breed (or think they do!) then just simply preface your response with 'then you know they have wonderful natures', etc.

If you are directly asked about price be honest and explain that it is an almost impossible question to answer because there are so many different prices in a single litter of piglets – individual prices vary depending on age, markings, conformation, sex and whether the piglet is for breeding, fattening or just for a 'pet'. In any litter of ten piglets there are probably about five or six different prices depending on a combination of all these factors in addition to whether or not the pig is to be registered. By far the best first step is to arrange a 'no obligation' visit for your potential customer.

Finish your telephone conversation by reaffirming that you will send further information with directions when you confirm the date and approximate time. During all your time on the telephone avoid 'selling' and just talk about your piglets, how delightful they are and how you will give the caller all the 'help' you can no matter what their pig keeping experience might be. Above all show a genuine interest in what the caller says, their needs and their problems as well as in them as a person. It is not sycophantic to look for people's good points and show an interest in them. All too often people are so wrapped up in themselves and their problems that they do not really listen to what other people have to say yet you will not find anyone who is successful in any business who is not a good listener.

SENDING ADVANCE INFORMATION

The information that you send out to those people who respond to your advertisement or make any enquiry about the piglets you have available should contain as much detail as possible. This information can be sent as an email attachment or by post but obviously the quicker you can send the information the better. It is at the point of making the enquiry that the prospective new pig keeper is at his or her most enthusiastic so the speedier your response then the greater will be your chance of securing that all-important visit to see your piglets.

If you respond by email you can include the following:

▶ *Directions on how to get to you.*
▶ *A few digital photographs of mother and piglets.*
▶ *A note thanking them for their enquiry and including some help and advice based on the information you acquired during your telephone conversation.*
▶ *Details of suitable websites that they might find interesting such as the websites of the particular breed society, the BPA website and, of course, your own website if you have one.*
▶ *Information about you and what you do in order that people do not feel you are a complete stranger when they arrive.*

If customers do not have access to the internet or do not have their own email address you can send them the above information by post and include a breed leaflet that you can get from the breed society.

Supplying all of this information does require a little extra effort but the added returns are worth it.

Advertising

Always advertise early. If you leave advertising your piglets until after they are weaned (8–10 weeks old) you may not get an actual visit from anyone interested for two or three weeks after that. The process of getting your advertisement accepted, people responding and then finding a mutually convenient time to visit can often take much longer than you might realize so place your advertisement either on the day they are born or up to two weeks prior to the event. Piglets are at their most appealing when they are young – when they reach 12 or 14 weeks of age they do not always have the same 'Aaah!' factor. If you leave your advertising too late then you may end up:

1 *feeding longer than you would want to and every extra day of feeding makes the cost effectiveness of rearing your first litter less impressive;*
2 *getting worried and concerned that they might not sell at all;*
3 *feeling pressured to sell at any price because they are growing so quickly and costing you money; and ...*
4 *taking visitors to see pigs that are no longer piglets and are less appealing. A pig at four months old only has two months to go before pork weight and will not create the same 'I must have' response as very young piglets.*

Your aim should be to have all you want to sell sold by the time they are six weeks old. Customers can collect at six to nine weeks depending on your preference and your own particular weaning programme. Provided your piglets are eating solid food well then weaning before eight weeks will do them no harm and could well prevent their mother's teats from being bitten unnecessarily by the piglets' sharp little teeth.

Once you become an established breeder you will no doubt have a number of enquiries for piglets all the year round and you may even find that your 'waiting list' of customers is more than adequate to ensure that you have outlets for your weaners even before they are born. Although you will never know the number of piglets you have available until they are born it is better to have more customers than piglets rather than more piglets than customers.

Insight: When a sale is not a sale

When I first started fattening pigs for the table I had nine people who had promised faithfully that they would have half a pig (butchered, wrapped, boxed and ready for the freezer) when they were ready. In the end only two kept their word. Remember an order is not an order until you have a deposit. No deposit – no order.

If people know you are genuinely enthusiastic about the breed yourself they will come and look. Your enthusiasm will be contagious. This gives them added confidence in what they are about to do and there is never any difficulty getting them to make a commitment by leaving a deposit. Your new customers can then come back a few weeks later to collect, once they have applied for a CPH if necessary and you have organized the required Movement licence.

If you plan ahead and aim to sell your weaners at realistic cost-effective prices you will certainly have a much better advertising budget for your litter than if you leave your pricing policy to float in the wind. You will also have more feed to provide if you have to keep your piglets and sell them as porkers because you were ill-prepared. Build an extra amount for advertising into your pricing because just a little extra per piglet gives your advertising budget a big boost.

If you have not yet built up a waiting list of customers then start by advertising locally (it makes travelling easier) in your local evening and weekly paper and don't be mean with copy. Do advertise for more than one day or week and advertise on consecutive days. Many local evening newspapers offer a special 'four nights for the price of three' deal because they know that it is often the third or fourth reading of an advert that prompts the best response. The first advert appears to just sow the seed and acts as a prompt while day two is often just a reminder that your piglets are for sale. Day three is the real attention grabber and it is often not until day four that you start to get the best telephone replies. Remember also that many people may not buy the local evening paper every night but simply choose to buy two or three times a week on impulse, so if you just advertise for one night you might miss these readers. By achieving a better price for your stock you will have an increased advertising budget and you can afford to advertise more. Many local papers also have a special 'farming or countryside' night so make sure you include that day in your advertising schedule.

Look at local evening and weekly papers in your neighbouring town or county as they can produce good results as well – don't restrict your market place – 60 minutes away in most directions lies a whole group of new potential customers. With this in mind, you could even think about developing your own website to reach even more potential customers.

The visit itself

If you are going to rely on your pigs to do the selling for you then they will need a little help from you when your potential customers pay their visit.

A good, clean natural environment is very important. If your stock looks as if it is loved and cared for it will support your own enthusiasm. If your pigs are easy to handle and react as if you have cared for them it will speak volumes to your prospective customers.

Make sure their bedding is clean (it should be anyway) and that the pigs can be seen easily. If you keep your pigs running naturally outside then remind your visitors to bring a pair of Wellington boots. If the weather is fine and dry then make sure you take a treat to attract your pigs – there is no need to shut them up as young piglets look even better running around naturally outside. If it is wet, try to show off your stock in a shed or covered situation that allows your visitors to have a close look – interaction between your piglets and customers is really important.

Show how docile and friendly the breed is. If your pigs have been well handled and treated properly they will 'perform' exactly as you want them to. Let your customers touch as well as look because your young piglets will be silky, soft, affectionate animals and your visitors may not appreciate how nice they really are until they touch them. You will hopefully have got your new stock used to being picked up by this time and their mothers will also be used to any squealing there might be and will simply ignore the noise.

It is polite and friendly to offer a cup of tea or something similar. The effort is worth it and it also gives you time to do the paperwork. You want to get a deposit (at least half the sale price), although a lot of people will pay in full. You need to agree a collection date and organize registration where applicable.

THE DEPOSIT

If your piglets have not yet been weaned, a deposit to secure any order is essential. Make sure that if the piglets are not being taken and paid for immediately that you get at least a 50 per cent deposit. This is because of the following:

▶ *A deposit secures the order and is a commitment by your new customer.*
▶ *A deposit helps your own cash flow by ensuring that the new owner of your piglets is paying for future feed – not you!*
▶ *Deposits on either a piglet or on a meat order helps make the end product look cheaper when they come to pay the balance on collection.*

Livestock markets and speciality sales

If you are left with older stock, have breeding sows you wish to move on or a registered boar you wish to change then you may consider markets, minority breed sales and auctions. While they serve a purpose and should never be overlooked it is always better to be totally in control yourself and if you can avoid the necessity of paying someone else to sell your stock then so much the better.

The disadvantages of selling by auction are as follows:

1 *The bid price you are likely to get could be very low unless you are extremely lucky and have two or three serious buyers all interested in your pigs.*
2 *Transport costs, your time (even if it is a nice day out), auctioneer's charges and commissions all mount up and eat into any bid you might accept. Most auctioneers don't pay out on the day because they need to make sure that they have the*

money from the buyer before they pay you. You may have to wait 14–30 days in some cases.

3 *The dealer 'vultures' are always on the look out for 'something for nothing' at local minority and rare breed sales as well as at local markets. They see coloured breeds of pigs (whether registered or not) as a natural target and will buy your well loved pig at the cheapest price possible only to sell as meat the next day.*

4 *The 'white pigs only' culture rules at most local markets. Even the prices paid at market for white pigs is uneconomically low so what chance will your rare breed, naturally reared or coloured pigs have?*

5 *There are always more 'sellers' and 'watchers' than 'buyers' at most minority breed sales.*

Having said all this I do accept that there may be times when you may want to (or even have to) take your pigs to a minority or rare breed sale, perhaps (at worst) to your local pig market. If you do, please bear in mind the following:

1 *Select a sale that is widely advertised within the world of pigs, smallholders and rare breeds. Such sales are held annually by the BPA, the RBST and the Wales and Border Counties Pig Breeders' Association. You will be able to look at some excellent stock at these sales and will also make many useful contacts. These sales are run by specialists for specialists and are well worth a visit even if you have no entries yourself.*

2 *Your pigs tell everyone at the sale about you and not just your pigs. If your stock is in poor condition, badly prepared or presented as a 'second-rate' pig you will only get 'second-rate' bids – if you get any at all! Worse still, everyone who sees your stock will know you are a poor pig keeper and word soon gets round. To take such stock to a sale is an insult to the breed, a bad advertisement for you, for the breed and everyone involved in promoting rare, minority and naturally reared pigs. If you can't be proud of your stock keep away from the exposure of sales and markets.*

3 *Set a reserve and stick to it. Don't sell at any price. Doing so only drags down the overall price and status of the breed and*

gives nobody satisfaction – least of all you. If you plan ahead with your sales, breeding and marketing strategy you should be able to avoid the desperate 'must sell at any price' strategy.

4 *Be near your pigs as much as possible prior to the sale commencing. If you are near the pen then you can talk to prospective buyers, sell the benefits of the breed and give the background history of your stock. No need to be 'pushy', just friendly and helpful because you are likely to get more bids if prospective purchasers know your stock has been kept and reared in the kindest, best possible way.*

5 *If you are not happy with the price bid say 'no' and take your pigs home again. You are going home yourself anyway. It is better to go home with your pride intact to look for another way to sell your stock than it is to drive home dispirited, angry, without your pig (and friend) knowing that you should have said 'no'. And what little money is due will perhaps not arrive for 30 days or so anyway.*

6 *If you do sell your pigs successfully (and I hope you do) then make sure you speak to the buyer. Pass on as much information as you can about your stock, hand over any certificates and make sure you get the buyer's name, address and telephone number because you now have another contact who might buy pigs from you in the future, thus avoiding the sales and markets.*

7 *Always spend a little time talking to the auctioneer about your pigs and their background. If you have specific information write a note to give the auctioneer at sale time along with any pedigree certificates you might have.*

Very often I find that 'selling pigs' is the most difficult part for some breeders and as most small-scale pig keepers keep pigs mainly for their own pleasure this is understandable.

The Rare Breeds Survival Trust meat scheme

If you have tried all other avenues and still find yourself unable to find a suitable market for your pigs then this scheme may be an avenue to explore (for more information on the Rare Breeds Survival Trust, please see Chapter 11). Although it does have its shortcomings it also has much to commend it. It is certainly better than taking your pigs to market and you will also be contributing towards the RBST funds and saving the rare breeds. The price you receive depends on weight. The 'finishing unit' you take your pigs to has the right to refuse any stock that is not up to standard so it is not a 'dumping ground' for poor-quality stock.

If you are members of the Rare Breeds Survival Trust (and I recommend that you are!) then you will find all the information about 'finishing units' and 'accredited butchers' in their quarterly journal *The Ark*. The RBST will be able to advise you of your local 'finishing unit'. We are sure that they will give you all the help and advice on the scheme that you require.

The Rare Breeds Survival Trust is always looking for additional butchers to join the RBST meat scheme and also pig keepers with suitable premises, experience and a desire to play their part in helping save rare breeds to act as a 'finishing unit'. To get more information and to discuss the potential opportunities contact the RBST direct.

Final thoughts

Never doubt that rare, traditional and naturally reared pigs are good value. If you ever doubt that your quality pigs, which are so 'special' to you, deserve to be priced accordingly then just think of the following:

- *The time and effort you have spent rearing them.*
- *The amount of pleasure and enjoyment your pigs have given you and will give their new owners also. They are intelligent and very endearing animals.*
- *The price currently paid for other pedigree (pure bred) animals. Check out the current prices for an eight week*

old Persian kitten, a newly weaned King Charles Spaniel and llamas or alpacas. It is a fact that your rare breed, traditional or naturally reared pigs and piglets are extremely competitively priced!

If you love pigs and pig keeping then you must make sure that they do not eat you out of house and home. Be sensible, be realistic, be cost effective and above all be professional in the way you sell and market your pigs and your meat. Give people your time and act courteously when they come to see your pigs or collect their meat and you should be rewarded with more sales and a much more enjoyable experience.

10 THINGS TO REMEMBER

1   *Never sell on price always on quality.*

2   *Do not sell pork, bacon, gammon, half pigs or whole pigs or you will fail. Sell a 'Mouth-watering eating experience', 'Crackling that crackles like it used to', 'Joints that don't shrink in the oven' and 'Bacon that doesn't drown in the pan'.*

3   *Your first sale is <u>not</u> the most important – it is the repeat orders and 'word of mouth' orders that matter most.*

4   *Any idiot can give things away or sell at a loss.*

5   *Always take a deposit. A sale is not a sale until it is collected and paid for.*

6   *If friends and family want 'something for nothing' when you have done all the work then what sort of friends and family are they really?*

7   *When selling use the 'four-letter word' 'H-E-L-P' as much as possible. You must do everything to show your customers that whatever they get from you (piglets or meat) is going to benefit, or help, them in one way or another.*

8   *Take advantage of the 'Aaaah!' factor and advertise early so your customers see your piglets when they are just a week or two old. The optimum time to advertise is two weeks before they are born.*

9   *Remember it is harder to answer an advert than place it so make sure your telephone skills are at their very best.*

10 *When you have sold all your piglets or all your meat then you should have a glass of 'red or white medicine' (the sort that comes from Chile, Australia, France, etc.) because the whole operation has been a success and this is down to only one person – you! Cheers!*

11

Who's who in the pig world

In this chapter you will learn:
* **useful contacts in the pig world.**

The most important website to keep in your 'favourites' file is www.defra.gov.uk. Changes in the law and rules as far pig keeping, feeding, identification and transporting pigs is concerned change regularly so this is the most important website if you want to keep up with what is happening and current legislation.

The British Pig Association

Although pigs have been part of our lives for thousands of years with the wild boar being the origin of the majority of our domesticated pigs it was not until very much later that the breeds as we know them now were clearly established.

Without a controlling body that would set and monitor breed standards, verify and register pedigree stock of a particular breed as well as record all notifications of births by means of a well-controlled Herd book there could be no specific breeds as we know them today.

The definitive date for the establishment of breeds as we know them is generally accepted as being 1884 – this being the date the British Pig Association was formed. Although prior to this date there had been a number of Herd or Stud books maintained by individuals there had been no universally accepted standards set and agreed for each individual's favourite pig. This resulted in such anomalies as several breeds of similar appearance and description being known by different names. Formed at the instigation of Mr Sanders Spencer, who was appointed Honorary Secretary of the National Pig Breeders' Association, in order in part to establish clear breed criteria, the first adopted breeds were the Large White, Middle White and Tamworth. Within the first 12 months the first Herd book was produced and this also included the Berkshire, Small White and Blacks. Points for each breed were gradually established covering colour, markings, the shape of the face, conformation and type of ears (lop, semi-lop or pricked). The breeds as we know them were established by man and not by nature, in order to meet and fill specific needs or requirements and over the years each breed has been gradually 'improved' by selective breeding.

If you are serious about breeding any of the breeds currently controlled by the BPA, maintaining breed standards and thereby contributing to the growth and development of a particular breed then you must join the British Pig Association. It is only by being a member of the association that you can register any pure bred stock you might breed and thus maintain their pedigree status. You cannot breed pedigree pigs that can be registered unless both the sow and the boar are both registered and all the necessary paperwork is in order.

Wales and Border Counties Pig Breeders' Association

If you live in Wales or anywhere in the bordering counties you will find that this association will take you under its wing as a newcomer to the world of pigs. Although sounding like a sibling of the BPA there is no connection whatsoever and the WBCPBA has no regulatory authority as far as registrations, birth notifications or other such matters are concerned. Despite this, the association performs a very worthwhile function by both promoting pigs and pig keeping in an ever-widening geographical area. You will find the WBCPBA in evidence at all the major Welsh shows as well as at the Smallholders' Show held annually at Builth Wells where they hold special pig keeping events especially designed for people new to the world of pig keeping.

Breed societies and clubs

There is a club or society associated with most breeds and if you have an interest in a specific breed they offer a very useful starting point for any new pig keeper. Most publish a regular newsletter for members, hold an annual general meeting and in many cases organize special social events. As a new pig keeper you will be warmly welcomed and given lots of help and advice. It is a good idea to look on the internet for the various websites associated with the various clubs and societies because they will give up-to-date information including the name, telephone number and contact details of the various committee members. If you cannot find a website for a particular breed then a call to the BPA or the RBST will result in you obtaining the name, address and telephone number of the breed secretary. The clubs or societies you may want to explore are as follows:

- ▶ *Berkshire Pig Breeders' Club.*
- ▶ *British Lop Society.*
- ▶ *British Saddleback Breeders' Club.*
- ▶ *Gloucestershire Old Spots Breeders' Club.*
- ▶ *Kune-Kune Society.*
- ▶ *Large Black Breeders' Club.*
- ▶ *Middle White Breeders' Club.*
- ▶ *Oxford Sandy & Black Club.*
- ▶ *Tamworth Breeders' Club.*

Rare Breeds Survival Trust

Last but by no means least is the body that anyone interested in saving our rare and endangered breeds should support. The Rare Breeds Survival Trust was established in 1973 in order to protect and save our rare and endangered farm animals and livestock. Thanks in large measure to their invaluable work over the intervening years we have not seen the loss of any of our farmyard breeds. Just prior to the formation of the RBST we saw the demise of our Lincolnshire Curly Coat pig in 1972 and this is something that should make all pig lovers sit up and think – how could we allow such a magnificent farm animal to become extinct in such recent times?

Glossary

AHDO The Animal Health Divisional Office in your area.

AI Artificial insemination.

AML Animal Movement licence. You must have one of these issued to you by the person who sells you your first pigs before you can move them to your property. This must be given to you when you collect your pigs. This is a self-issue document that must accompany any pig leaving or entering your property.

boar A male pig that has not been castrated.

boar taint A term used to describe the pungent smell while cooking, and the strong taste while eating, that is usually present if you produce meat from an entire boar over the age of about seven months that has not been castrated. Some very knowledgeable veterinary surgeons as well as established pig keepers say the pork will still be fine up to ten months of age provided he has not 'worked'. I have produced excellent pork with no sign of boar taint up to eight months of age. Never use an entire boar to try to produce bacon, ham, gammon or cured meat of any kind. If you do, you will wish you had not!

BPA The British Pig Association.

brimming This is when a female pig comes into season and is ready to accept the boar. This will happen every three weeks and will last for three days. Signs of 'brimming' include: possible change of temperament, less interest shown in food, swelling of the vulva. When pressure is applied to the back a 'brimming' pig will stand rigidly still and adopt the stance she would take when accepting the boar.

close to profit An expression used to indicate that a sow or gilt is due to farrow shortly. The nearer she is to the farrowing date then the closer she is to profit. This is based on the assumption that she will produce a good sized litter which, when sold, will produce a good return for the breeder, well exceeding his costs of feeding and keeping her thus resulting in a good 'profit'.

CPH County Parish Holding Number. This is something you must have before you can keep pigs. This is acquired by contacting your local RPA, DEFRA or Environmental Health (Animal Health section) office. Ultimately you will receive your CPH via the RPA. This is your starting point for keeping pigs.

creep The secure area where piglets can feed.

creep feeding The term used to describe the way in which piglets are fed in a secure area (the 'creep') that can be easily accessed by the piglets but not by the sow.

crush A means of containing the pig in a confined area so that you can easily administer drugs, examine or treat the pig, trim its nails (clays) or administer semen using an artificial insemination kit.

DEFRA The Department for Environment, Food and Rural Affairs.

farrow or farrowing The process of a pig giving birth to piglets.

finishing unit A farm that specializes in fattening pigs for pork and/or bacon but may not breed pigs themselves. They will buy young piglets as soon as they are weaned and feed them up to the appropriate fattened weight, usually in intense conditions.

gilt Most people will accept that a 'gilt' is a maiden female who becomes a 'sow' after she has had her first litter but officially she remains a 'gilt' until she is 'in-pig' for the second time.

herd number An exclusive number issued to you by your local Trading Standards Office (Animal Health) once you have been allocated your CPH number. This is used when required for marking and identifying your pigs to adhere to government regulations.

hog or castrate A castrated male pig.

in-pig This is the term used for a sow or gilt that is known to be expecting a litter of piglets.

litter A group of piglets from the same mother up to about eight weeks of age.

long meadow An expression used in the past by cottagers and smallholders to describe the grass verges along the lanes and roads around where they lived. Because they owned little land and grazing themselves, they would often put the pig in a harness and take it to the 'long meadow' where it could happily graze all day tethered to an iron stake. It was more likely that the children of the household would take a 'pig' to school with them rather than a 'lamb', tethering it on the way to school and collecting it on the way home.

rattling The piglet that requires the greatest assistance to survive (see Runt).

RBST The Rare Breeds Survival Trust.

RPA The Rural Payments Agency.

runt The smallest piglet in the litter usually requiring a great deal of assistance from the pig keeper so that it can survive those first weeks of life. The runt often has to be bottle fed as its fellow siblings will often bully it.

scour/scouring Diarrhoea.

slap marker A board with a number of tattoo pins fixed permanently to it in a form that represents your own herd number. When these pins are dipped in a tattoo paste and then slapped hard against the shoulder of a pig an impression of your herd number is left on the pig.

sow An adult female pig that has had at least one litter of piglets.

underline The exact number of teats on a pig whether male or female. Added emphasis is placed on how they are spaced as well as how even they are – a very important factor to take into account when breeding and registering pure bred pigs.

WBCPBA The Wales and Border Counties Pig Breeders' Association. Although unconnected directly with the BPA, the WBCPBA is an association that does a great deal of excellent work helping new and first-time pig keepers who live in and around Wales and the border counties.

weaner A young piglet that has grown strong enough to be separated from its mother and is no longer suckling as it is eating solids. Normally at about seven to ten weeks of age, although commercially much earlier.

withdrawal time The number of days or weeks that must elapse between you or the vet administering a particular drug or medicine (including wormer) and that pig entering the food chain.

working boar An adult male that is used specifically for breeding.

Taking it further

Government departments and national bodies

ADAS – Pig Research Unit – 01553 828621 – www.adas.co.uk

British Pig Association (BPA) – 01223 845100 –
www.britishpigs.org.uk

British Pig Association AI centre – Deerpark Pedigree Pigs –
02879 386287 – www.deerpark-pigs.com

British Veterinary Association – www.bva.co.uk

Compassion in World Farming – 01730 264208 – www.ciwf.org.uk

Department for Environment, Food and Rural Affairs (DEFRA) –
0845 9 33 55 77 – www.defra.gov.uk

Irish Organic Farmers' and Growers' Association – 00 353 506 32563

National Organic Centre Wales – 01970 622100 –
www.organic.aber.ac.uk

National Pig Association – 020 7331 7650 – www.npa-uk.org.uk

Organic Farmers and Growers Ltd – 01743 440512 –
www.organicfarmers.uk.com

Organic Food Federation – 01760 720444 – www.orgfoodfed.com

Organic Trust Limited – 00 353 185 30271

Pig Paradise – one-day, rare breed pig keeping courses – 01380 723113 – www.pigparadise.com

Piggywigs Farm – small-scale pig keeping courses and rare breed pig visits for schools – 01380 722522 – www.piggywigsfarm.co.uk

Pig Veterinary Society – www.pigvetsoc.org.uk

Rare Breeds Survival Trust – 0247 669 6551 – www.rbst.org.uk

Rural Payments Agency – 0845 603 7777 – www.rpa.gov.uk

Scottish Organic Producers' Association – 0131 335 6606 – www.sopa.org.uk

Soil Association – 0117 914 2400 – www.soilassociation.org

The Pig Site – www.thepigsite.com

Wales & Border Counties Pig Breeders Association – www.pigsonline.org.uk

---

Breed clubs and societies

Berkshire Pig Breeders' Club – 01386 840375 – www.berkshirepigs.org.uk enquiries @berkshirepigs.org.uk

British Kune-Kune Pig Society – 01736 810519 – www.britishkunekunepigsociety.co.uk

British Lop Pig Society – 01767 315926 – email: balshampigs@tiscali.co.uk

British Saddleback Breeders' Club – 01285 860229 – email: saddlebackpig@aol.com

Gloucestershire Old Spot Breeders' Club – 07768 368053 – www.oldspots.com

Large Black Breeders' Club – 01694 731318 – www.largeblackpigs.co.uk

Middle White Breeders' Club – 01753 654166 – email: miranda@middlewhites.freeserve.co.uk

Oxford Sandy & Black Club – 01905 821828 – www.oxfordsandypigs.co.uk

Tamworth Breeders' Club – 020 8441 8961 – email: sarahbrickell@aol.com

---

Suppliers of pig equipment

Deer Park Pedigree Pigs (Robert Overend) – 028 793 86287 – suppliers of artificial insemination kits and rare breed pig semen.

Hanco Pig Supplies – 01432 860518 – national supplies.

Homefield Farm Services Ltd. – 01904 491490 – sole UK agents for the slap-shot injecting aid.

Ifor Williams Trailers – 01490 412626 – www.iwt.co.uk

Pig Paradise (traditional wooden arks) – 01785 280791 – www.pigparadise.com

Ritchey Tagg (tagging and livestock products) – 01765 689541 – www.ritchey.co.uk

Slap-shot injecting aid – www.slapshot-flex-vac.com

Magazines and periodicals

Country Smallholding (monthly) – www.countrysmallholding.com

Farmers' Guardian (weekly) – www.farmersguardian.com

Farmers' Weekly – www.fwi.co.uk

NFU Countryside (monthly membership magazine) – www.countrysideonline.co.uk

Smallholder (monthly) – www.smallholder.co.uk

Books

A Guide for New Keepers – Pigs – DEFRA.

Code of Recommendations for the Welfare of Livestock – Pigs – DEFRA.

Sites Suitable for Outdoor Pig Farming – DEFRA.

Fearnley-Whittingstall, H. (2004) *The River Cottage Meat Book*, Hodder & Stoughton.

Harris, C. (2005) *A Guide to Traditional Pig Keeping*, Farming Books and Videos.

Porter, V. (1993) *A Handbook to the Breeds of the World*, Cornell University Press.

Watson, L. (2005) *The Whole Hog*, Profile Books Ltd.

A Taste of Pork, available from the Wales and Border Counties Pig Breeders' Association – www.pigsonline.org.uk

Index